Exploring Science

Exploring Science

Experiments with electricity, magnetism, heat and light for primary and middle schools

by

T J Harvey

Lecturer in Education
University of Bath School of Education

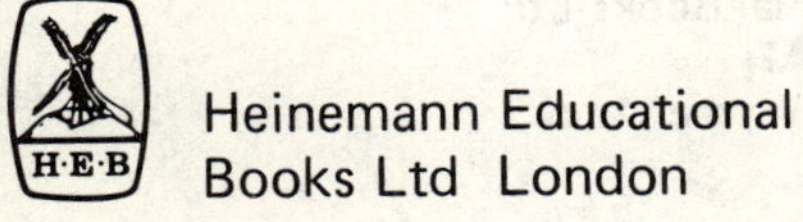

Heinemann Educational
Books Ltd London

Heinemann Educational Books Ltd
LONDON EDINBURGH MELBOURNE
JOHANNESBURG AUCKLAND
TORONTO SINGAPORE
IBADAN HONG KONG LUSAKA
NAIROBI NEW DELHI KUALA LUMPUR

ISBN 0 435 80608 4

Published by Heinemann Educational Books Ltd
48 Charles Street, London W1X 8AH
Printed in Great Britain by Cox & Wyman Ltd
London, Fakenham and Reading

Introduction

This book, containing thirty-two elementary experiments in electricity, electrostatics, magnetism, heat and light is intended mainly for the non-science trained teacher, but it is hoped that experienced science teachers will find it useful and stimulating. The experiments are set out as separate entities, with the main aim being to contain sufficient information to cater for the teacher teaching the topic for the first time.

In addition to the information relevant to each experiment, there are sections devoted to the availability of components and the construction of the apparatus. In providing this information it was hoped that schools having very little money to spend would still be able to teach some science, and more important if the children could be encouraged to build the apparatus, it would provide an integration between science and craft.

If these are completely new topics for the teacher, then it may well be advisable to teach the experiments, for the first time, rather formally, possibly in the manner suggested in the text, but it is hoped that in succeeding years the teacher will incorporate them into his normal teaching method. It is not the aim of this book to suggest or recommend any particular teaching method.

A teacher may, quite understandably, feel that he does not wish to teach all of these experiments. This is quite acceptable, since although they follow a logical pattern, they are not intended to be so interdependent one upon the other that they form a rigid course. It is hoped that teachers will choose those that they feel appropriate and integrate them into their own scheme of work, since essentially this book should be regarded merely as a source book for the teacher, with special reference to those completely new to the topics.

Contents

List of experiments

1. Simple experiments with charged bodies.
2. The forces between charged bodies and other bodies.
3. *Perspex-like* and *acetate-like* charges.
4. A simple electrical circuit.
5. One use of a switch.
6. To investigate the heating effect of an electric current.
7. To investigate the electrical conductivity of solids.
8. To investigate the electrical conductivity of liquids.
9. Series and parallel connections of bulbs.
10. Magnetic properties of materials.
11. Properties of the poles of a magnet.
12. Making magnets.
13. Making a compass.
14. Plotting magnetic fields.
15. To show that an electric current produces magnetism.
16. To make an electro-magnet.
17. To show that an electric current can produce movement.
18. To construct a simple electric motor.
19. Experiments with electric motors.
20. Producing electricity.
21. The simple dynamo.
22. Expansion of solids, liquids and gases.
23. The thermometer.
24. To make a simple thermometer.
25. To show convection currents in gases.
26. Convection currents in liquids.
27. Heat transfer by conduction and radiation.
28. The production of coloured light.
29. To mix coloured lights.
30. Light travels in straight lines.
31. Reflection by a plane mirror.
32. The bending of light.

Chapter One
Electrostatics, Magnetism and Current Electricity

Section One

Experiment 1
Simple experiments with charged bodies

Objectives

a To show that certain objects will attract other objects, *when they have been rubbed.*

b To show that this does not apply to metals.

Apparatus per group

Selection of plastics, pieces of metal e.g. copper, aluminium and iron, tissue paper, plastic comb, drawing pin, wooden ruler, balloon.

Procedure

i This first experiment is probably best performed by the teacher in front of the class as a demonstration.

Hold an unrubbed piece of plastic near to a jet of water coming out of a tap. The jet of water should be as fine as possible. Note that nothing happens to the jet of water. Rub the plastic with a piece of cloth or woollen object and again hold it near to the jet of water. The water should be bent towards the plastic i.e. the rubbed plastic has attracted the water jet.

Repeat the procedure with a piece of metal showing that in this case the rubbing has no effect.

The children could then try with a variety of objects to see which when rubbed will attract the water, and which will not.

ii Give out small pieces of tissue paper to the class and ask them to bring a plastic comb near to the paper — nothing should happen. Ask them how they can make the comb attract the paper — this can be done by combing their hair. They can then check that this only applies to plastic since rubbing a piece of metal on their hair will not cause the metal to attract the tissue paper.

The children can extend this experiment by finding the largest piece of paper that their rubbed comb will lift.

iii Give out a ruler and a drawing pin and ask the class to balance the ruler upon the drawing pin so that the ruler is free to rotate. Tell them to bring a piece of unrubbed plastic near to one end of the ruler — nothing should happen. Ask how they can make the ruler move without touching it or blowing it.

This can be achieved by rubbing a piece of plastic and bringing it near to one end of the ruler when it will attract the ruler causing it to rotate.

This can be repeated using other plastics and metals.

iv Give out blown-up balloons and ask the children how they can be fastened to the wall without using glue, sellotape or string.

If the balloon is rubbed with cloth or wool it will easily stick to the wall, whereas an unrubbed ballon will not stick.

Notes

1 A body that shows this attracting property when rubbed, we call a *charged* body.

2 A charged body can be discharged by touching it by hand in the regions where it has been rubbed to charge it. This should be shown experimentally at some stage.

3 All the apparatus used in Experiments 1, 2 and 3 must be kept dry, since if they become damp they quickly become discharged. Hence it is advisable to keep all the apparatus in a box on the radiator for at least 12 hours before use. The best results with these experiments are achieved on dry, warm days.

Experiment 2
The forces between charged bodies
and other bodies

Objectives

a To show that a charged body will attract an uncharged body.

b To show that two identically charged bodies will repel one another.

c To show that two charged bodies can attract one another.

Apparatus per group
Two balloons, length of cotton, selection of plastics.

Procedure
i This first experiment is probably best performed by the teacher in front of the class as a demonstration.

Attach a piece of cotton to a blown up balloon, and suspend the balloon so that it hangs freely and is not too near any other objects, for example a wall. Make sure that the balloon is discharged by stroking it all round with your hands.

Bring near to this balloon a second discharged blown-up balloon, and note that there is no force between them.

Charge the second balloon by rubbing it with cotton or woollen material, and bring it near to the suspended balloon. You should observe that the second balloon now attracts the suspended balloon, i.e. a charged body attracts an uncharged body − this was shown but probably not specifically stated in experiment 1.

Charge the suspended balloon by rubbing it, and bring near to it the charged second balloon. Since both balloons are the same, then presumably the charge on each will be the same, i.e. two *like* charges are now being brought together. You should observe that the two balloons *repel* each other i.e. *like* charges *repel*.

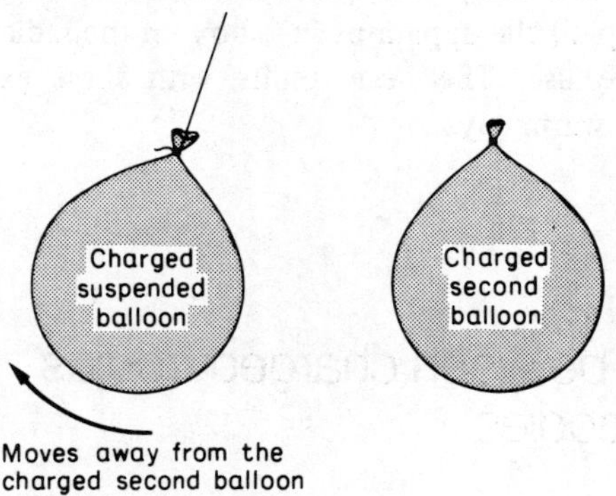

ii A selection of suspended charged balloons should be set up, the main point being that they should be reasonably clear of other objects, such as, for example, the wall, doors or windows.

The class should then bring near to the suspended charged balloons a selection of charged plastics, noting whether attraction or repulsion has occured.

Since some of the charged plastics repel the charged balloon, and some attract it, there would appear to be two different types of charges. Those that attract the balloon have a different charge from the balloon, i.e. *unlike* charges *attract*.

Consequently repulsion occurs between like charges, whereas attraction occurs between unlike charges, or between charged and uncharged bodies.

Notes

1 To obtain maximum effect the balloon must be rubbed all round when charging it.

2 Touching the suspended balloon will discharge the regions touched and hence reduce its effectiveness.

3 Hold the largest area of the plastic that has been rubbed nearest to the balloon.

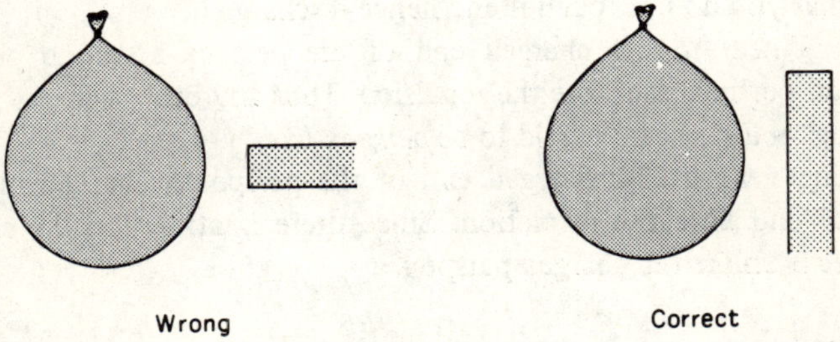

Experiment 3
Perspex-like and acetate-like charges

Objectives

a To draw up a list of those charged bodies similar to charged perspex and those similar to charged acetate.

Apparatus per group

Two pieces of perspex, two pieces of cellulose acetate, a paper sling with cotton, and a selection of other plastics or objects that can be charged by rubbing.

Procedure

i The results of Experiment 2 have shown that some charged bodies will attract one another. This is particularly true of perspex and acetate. If you do not have either perspex or acetate, then substitute two plastics that will attract each other when charged. These will then have different charges on them when rubbed and charged.

ii The first part of the experiment is probably best performed by the teacher in front of the class as a demonstration, but it can easily be performed by the children providing that they are given sufficient experimental instructions.

Suspend a paper sling by means of cotton (see apparatus construction, page 59) so that it hangs freely and is not too near any objects.

Charge one end of a piece of perspex and place it in the sling making sure that you do not touch it and hence discharge it.

Bring near to the charged end of the perspex a second piece of charged perspex and note the repulsion. Thus any other charged object causing repulsion can be said to be *perspex-like.*

Bring near to the charged end of the perspex a charged piece of acetate and note the attraction. Since there is attraction the charged acetate is unlike the charged perspex.

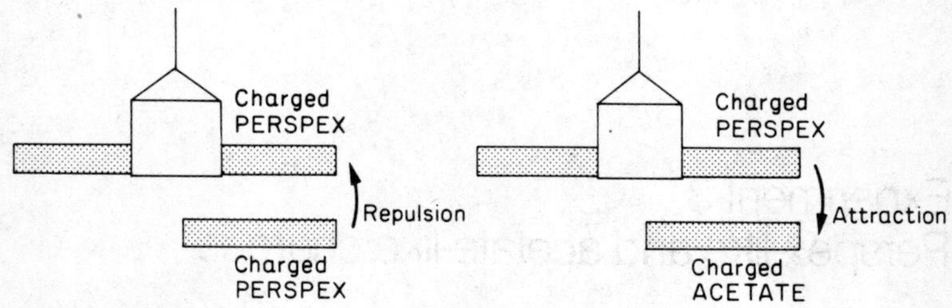

Replace the charged perspex in the sling with charged acetate and show that a second piece of charged acetate will cause repulsion whereas a piece of charged perspex will cause attraction.

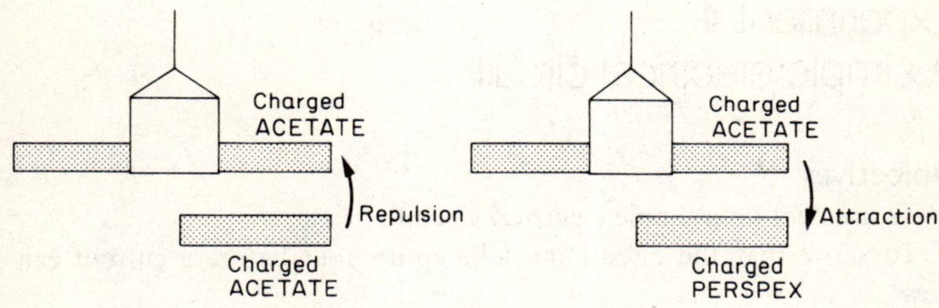

These experiments clearly show that we can classify charged substances as either *perspex-like* — if they repel perspex — or *acetate-like* — if they repel acetate.

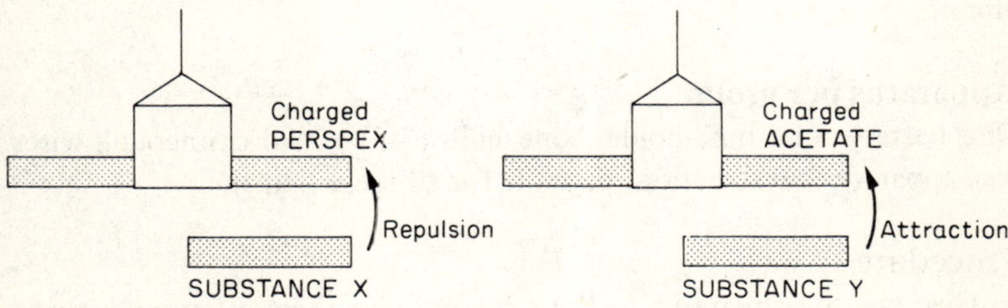

Substance X is thus *perspex-like*.   Substance Y is thus *acetate-like*.

iii Give out a paper sling with cotton attached, a piece of perspex and a piece of acetate to the class. They can now charge the acetate and place it in the sling and bring near to it a selection of charged plastics. Anything giving repulsion can be listed as *acetate-like*, anything giving attraction should be set on one side and tested later against a piece of charged perspex placed in the sling to confirm by repulsion that it is to be listed as *perspex-like*.

Notes

1 In this experiment it is the charges on the various charged bodies that we are trying to equate not the substances themselves.

2 For very obvious reasons it is impracticable to call the charges on the charged bodies as either *perspex-like* or *acetate-like*, and so we call them either *positive* (+) or *negative* (−).

Charged perspex is in fact + and charged acetate −.

Experiment 4
A simple electrical circuit

Objectives
a To connect up a simple electrical circuit.
b To show that the circuit must be continuous before a current can flow.
c To establish that the bulb can be used as a detector of electricity.
d To introduce the correct electrical symbols.
e To introduce the direction of current flow.
f To show that direction of current flow is not important in this circuit.

Apparatus per group
One battery, one bulb-holder, one bulb (3.5V), two connecting wires (see apparatus construction, page 60, for all these items).

Procedure
i Give out the apparatus to the class and give them a few minutes to investigate what they can find from it. *Keep an eye open to see that they do not connect the two terminals of the battery together* — this will quickly run the battery down.

If the class have not discovered how to light the bulb the following instructions should be sufficient.

"Connect up one terminal of the battery to one side of the bulb-holder, connect the other terminal of the battery to the other side of the bulb-holder".

To emphasize that the bulb only lights when the circuit is complete, disconnect and reconnect one of the connections several times.

Make sure that the clips connecting the wires to the battery terminals do not touch each other.

ii *Electrical Symbols*

The battery is —|⊢—

The longer vertical line being the positive (+ve) terminal.
The shorter vertical line being the negative (−ve) terminal.

Batteries will either have a '+' and a '−' on them or be marked 'RED' for +ve and 'BLACK' for −ve.

The bulb and bulb-holder is 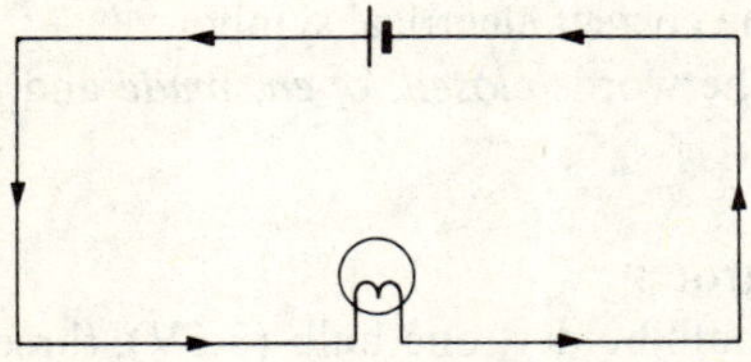

The connecting wire is never really shown. Consequently the circuit used was:

iii *Electrical Current Flow*

We say that the current flows from +ve to −ve i.e. as indicated by the arrow in the circuit diagram. This is only a convention and no proof of reasoning is necessary.

iv Get the children to reconnect the circuit and then reverse the connections to the bulb. They should see that the brightness of the bulb is the same in both cases, showing that the direction of current flow is not important in this circuit.

Notes

1 Introduce the use of the words *'shorting out'* the battery if the two terminals of the battery are joined directly with a single piece of wire.

2 The bulb and bulb-holder may be regarded as one electrical component.

3 The reason for establishing the correct electrical symbols is that diagrams should be correct electrical diagrams and not pictures. This speeds up the recording of the experiment and is the normal procedure in any further study of electricity.

Experiment 5
One use of a switch

Objectives
a To show one use of a switch.
b To introduce the correct electrical symbol.
c To introduce the words *closed, open, made* and *broken,* as used in electrical circuits.

Apparatus per group
One battery, one bulb-holder, one bulb (3.5V), three connecting wires, one tapping key.

Procedure
i Before beginning the experiment it is advisable to discuss the switch with the class.

The connections are made as indicated. The switch is *closed,* i.e. the electrical circuit *made* by pushing the metal down, and the switch is *opened* i.e. the electrical circuit *broken* by allowing the metal to rise up again. The construction of the switch is described in detail in the apparatus construction section.

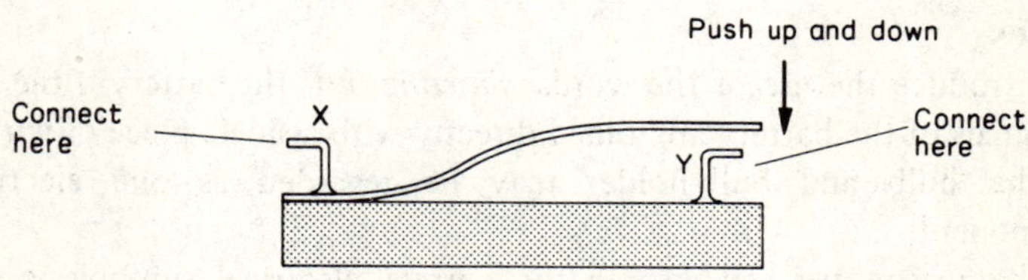

The electrical symbol for the key or switch is

This indicates that the key or switch is *open,* i.e. in this position there would be no electricity flowing around a circuit.

ii Give out the apparatus to the class and let them connect up the following circuit.

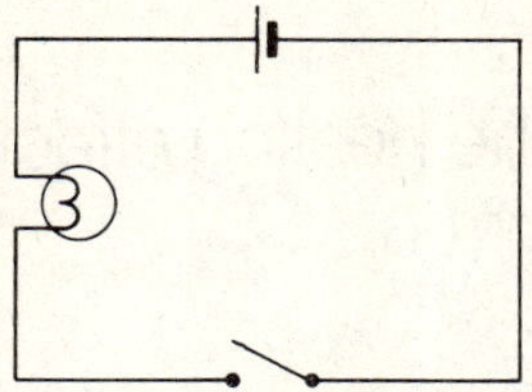

Care should be taken to see that the components are all connected into the circuit in the correct order and the right way round.

The correct way for the switch is —————•⟍•—————
Y   X

where X and Y are the connecting points indicated in the earlier diagram.

Let the class push down and release the metal on the tapping key. They should see the bulb light up and go out.

This then gives a method of communication using the MORSE CODE. Use long periods of "light on" for the dashes, and short periods of "light on" for the dots.

Let the children transmit their own name, messages etc.

Morse code

A • —	B — • • •	C — • — •
D — • •	E •	F • • — •
G — — •	H • • • •	I • •
J • — — —	K — • —	L • — • •
M — —	N — •	O — — —
P • — — •	Q — — • —	R • — •
S • • •	T —	U • • —
V • • • —	W • — —	X — • • —
Y — • — —	Z — — • •	

Notes

1 In this and all the circuits that follow it is not essential that the components are exactly in the same position as indicated in the circuit diagrams, but it is very good training for the children to follow the circuit diagram exactly. In later life if they try to build radios, etc. they must follow the diagrams exactly.

Experiment 6
To investigate the heating effect
of an electric current

Objectives
a To show that there is a heating effect associated with an electric current.
b To establish the mechanics of a fuse.

Apparatus per group
One battery, one bulb-holder, one bulb (2.5V), two pieces of connecting wire, the fuse tester (see apparatus construction, page 62), some steel wool.

Procedure
i Give out sufficient apparatus to set up the circuit below, making sure that you now use the 2.5V bulb.

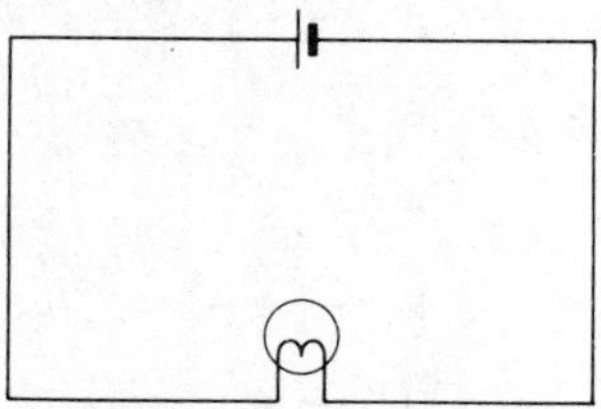

After the bulb has been on for about a minute ask the children to touch it with the back of their hands or to place the bulb on the inside of the upper lip. They should feel that the bulb is warm, thus showing that the electricity is producing both light and heat.

ii Discuss the electric fire which produces less light and more heat, and the domestic iron which produces no light and only heat.

iii Remove the bulb holder from the circuit and give the class some wire wool to put in its place. They should use about 10-20 strands of wire wool clamped in one crocodile clip touching the rest of the wire wool. They should observe smoke and sparks.

iv The fuse tester is now used to investigate the current passing through a single piece of wire wool. Issue the fuse tester and connect up the circuit thus:

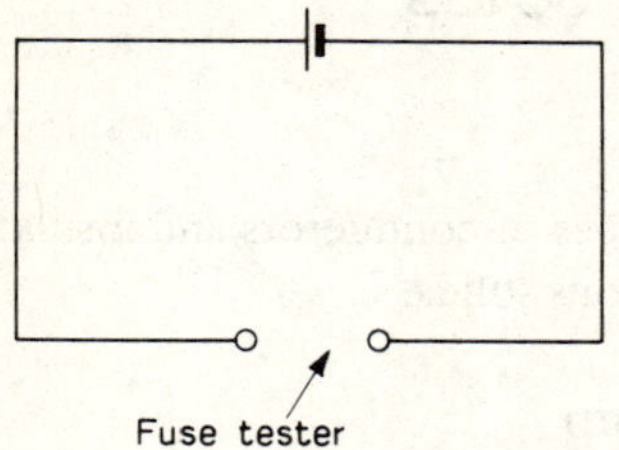

A single piece of wire wool should be held taut between the thumb and first finger of each hand. It should then be brought into contact with the two nails on the fuse tester. If the piece of wire wool is thick enough it should glow before melting, if it is too thin the melting process will occur too quickly. Consequently the children should be encouraged to try this with different pieces of various thicknesses.

N.B. It is advisable for the teacher to demonstrate this last technique *without* the battery connected.

If this procedure proves too difficult it is possible to wind the single piece of wool around the nails on the fuse tester.

Notes

This last experiment demonstrates the action of the fuse. It is important to establish the way in which the fuse works.

1 The electricity produces heat which causes the wire to get hot and glow.

2 The hot wire then melts and so breaks the electrical circuit.

Thus there are two process: **1** the wire glows, **2** then it melts.

In a household fuse these two actions occur almost simultaneously.

Experiment 7
To investigate the electrical
conductivity of solids

Objectives
a To establish the ideas of conductors and insulators.
b To investigate various solids.

Apparatus per group
One battery, one bulb-holder, one bulb (3.5V), three pieces of connecting
wire, emery paper.

A selection of solids: rusty nail, wood, paper, comb, various types of
plastic, string, wool, cotton, a pencil sharpened at both ends, various
types of metals.

Procedure
i Give out sufficient apparatus to set up the circuit below.

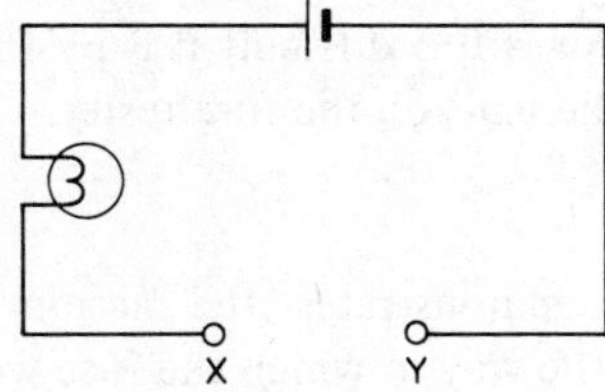

X and Y are the free ends of two pieces of connecting wire.

Join X and Y together to check that the circuit is all right and to
revise that current flow is detected by the bulb lighting up.

Introduce between X and Y selections of materials and see if the
light comes on or stays off.

Conductor – this will allow electricity to flow through it easily.

Insulator – this will NOT allow electricity to flow through it.

Since conductors will allow electricity to flow through them, when
placed between X and Y the light should come on; whereas insulators
will not allow the electricity to flow, hence preventing the light coming
on.

Two interesting cases are the rusty nail and the graphite centre of the pencil.

Rusty Nail

The rusty nail should not conduct, but on removing the rust with the emery paper it should then conduct. From this one should emphasize that rust is an insulator and that all electrical contacts must always be clean.

Pencil

The wood of the pencil will not conduct, but what may surprise the children is that the graphite centre will conduct — hence the necessity of sharpening both ends of the pencil.

Notes

1   You may find that the rusty nail appears to conduct even though it is rusty. This is probably because the teeth of the crocodile clips have bitten through the rust layer and reached the pure metal underneath. To overcome this it may be advisable to touch the ends of the rusty nail with the crocodile clips rather than clip them on.

Experiment 8
To investigate the electrical conductivity of liquids

Objectives

a   To investigate various liquids.

Apparatus per group

One battery, one bulb-holder, one bulb (3.5V), three pieces of connecting wire, the liquid tester (see apparatus construction, page 61).

Selection of liquids

Ink, paraffin, acid from a car battery or an accumulator, vinegar, washing-up liquid, orange squash. A bottle of pure water, or for the purpose of this experiment a bottle of tap water, labelled "PURE WATER".

Solutions to make up

Add fairly large quantities of the following solids to warm water: common cooking salt, copper sulphate crystals, washing soda, bath salts, magnesium sulphate. After the maximum amounts have dissolved, pour off the liquid into labelled bottles or containers.

Common cooking salt — label it "cooking salt dissolved in water".
Copper sulphate crystals — label it "copper sulphate dissolved in water".
Washing soda — label it "washing soda dissolved in water".
Bath salts — label it "bath salts dissolved in water".
Magensium sulphate — label it "magnesium sulphate dissolved in water".
(See apparatus construction, page 62, for further instructions).

Procedure

i   Give out sufficient apparatus to set up the circuit below.

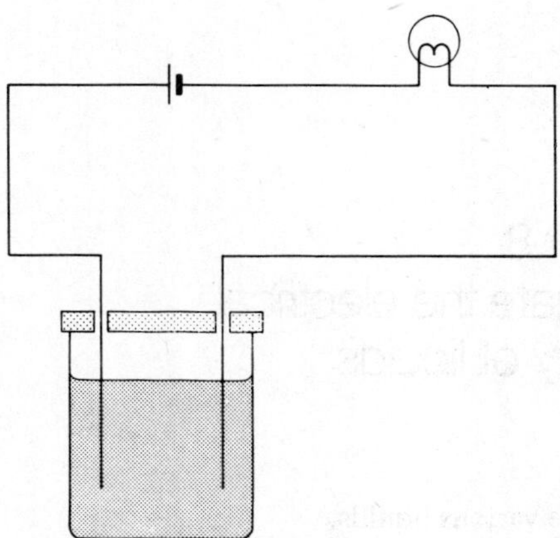

With no liquid present the bulb does not light up. Check that the circuit is all right by joining the two nails together and seeing that the light comes on.

Add sufficient of the pure water to cover the bottom three inches of the 6 inch nails. The light should still not come on.

Replace the water with one of the other solutions.

Repeat the procedure with the other liquids making sure that you clean out the container and the nails before using the next liquid.

Draw up a list of conductors and insulators.

Notes

1 It would appear from your results that water will in fact not conduct electricity. This apparatus is not very sensitive but strictly it is only pure water that will not conduct electricity. You would thus be advised not to let the children see you get the water from the tap since it is a good thing to point out that tap water is impure and will generally conduct electricity, thus they should not touch switches with wet hands. In bathrooms only pull-cord type switches are allowed because of the danger of electrocution.

2 It is important that the liquids are saturated, i.e. as much solid as possible has dissolved in the water.

3 Make sure that the nails have been wiped dry and clean before using another liquid.

4 Make sure that the bulb you use is rated 3.5V 0.15 Amp. This point is very important.

5 It may be necessary to shake the liquid tester slightly to get the light to come on. Most of the liquids that you have made up should be conductors.

6 It is best to finish up using the copper sulphate solution, since there is the additional effect of copper plating of the nails.

7 The correct name for the liquid tester is a voltameter, but since the voltameter is also used for other things it is reasonable for this experiment just to call it a liquid tester.

Experiment 9
Series and parallel connections of bulbs

Objectives
a   To establish the concept of series and parallel connections.

Apparatus per group
One battery, two bulb-holders, two bulbs (2.5V), four pieces of connecting wire.

Procedure
i   Give out the apparatus to each group and ask them to connect up this circuit:

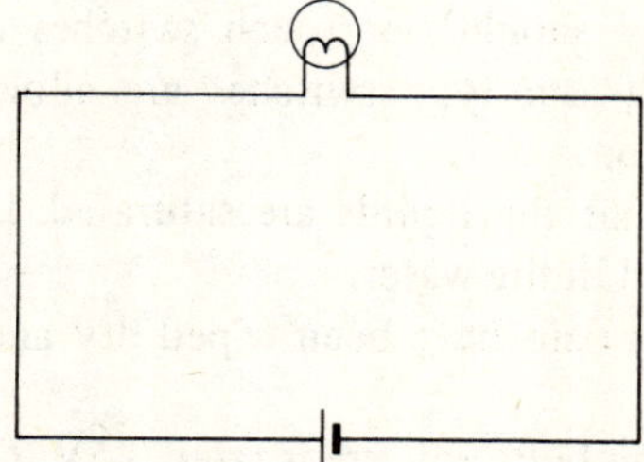

Note the brightness of the bulb.
Now ask the class to connect up two bulbs thus:

These two bulbs are connected in *series*.
Then ask the class to connect the battery to them thus:

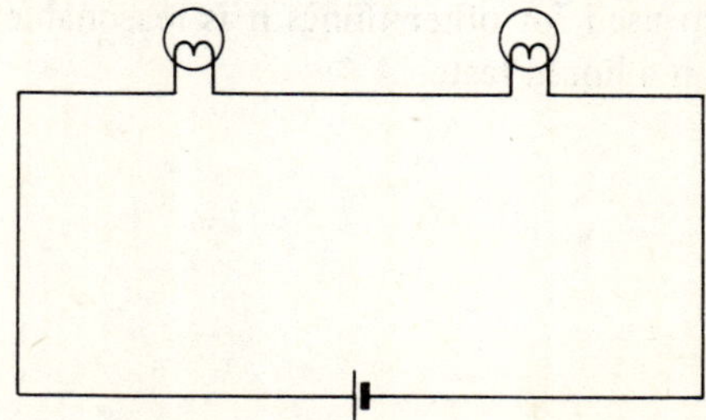

Note the brightness of the bulbs. They should be about equal brightness — this shows that the same amount of current is passing through both but they should be duller than in the single bulb connected across the battery at the beginning of the experiment. This then shows that the amount of current taken from the battery goes down when you have two bulbs connected in series.

Unscrew one of the bulbs and show that this puts both bulbs out. Repeat with the other bulb. This shows that the electricity passes through both bulbs one after the other and stopping it going through one stops it going through the other.

N.B. This is how Christmas tree bulbs are connected and thus if one goes the whole lot will go out. How do you then check which bulb is the one that has gone?

ii Now ask the class to connect two bulbs thus:

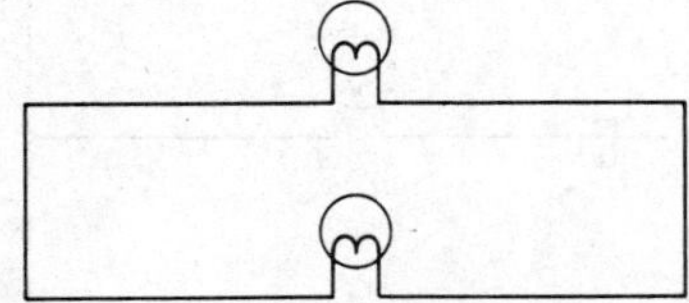

These are now connected in *parallel*.
The class should then connect the battery to them thus:

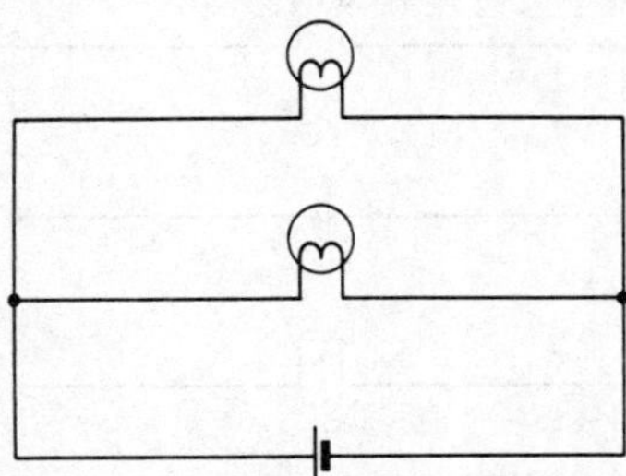

Note the brightness. They should both be about equal brightness showing that the same amount of electricity is passing through each. In addition they should be about the same brightness as the single bulb across the battery at the beginning of the experiment. This would indicate that two bulbs connected in parallel take twice as much current from the battery as the single bulb did. Thus two bulbs in parallel will run the battery down much quicker than two bulbs in series, but will give off much more light.

Unscrew one of the bulbs and see that the other one stays on and its brightness is unaffected. Repeat the procedure unscrewing the second bulb.

Thus another difference emerges, since in series connections one bulb going out causes all the others to go out which is not the case in parallel connections. Household lights and car lights are connected in parallel.

iii Small groups within the class can now combine and pool their apparatus to build up some of the circuits discussed in the following notes:

Notes
This circuit

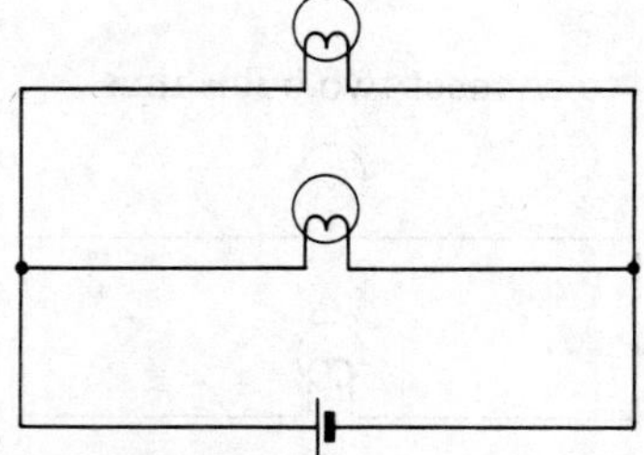

can be redrawn thus and is the same.

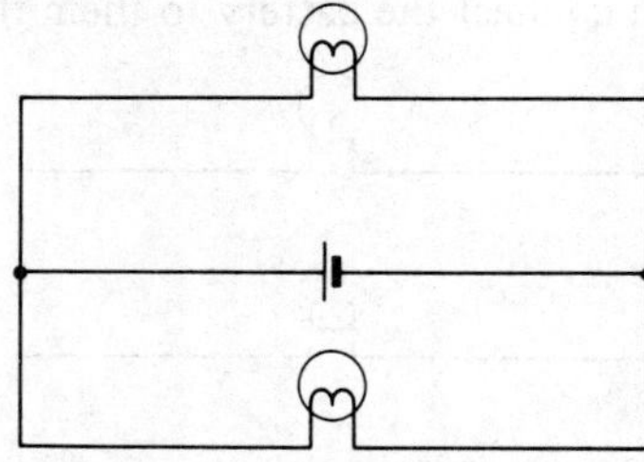

This circuit is also the same:

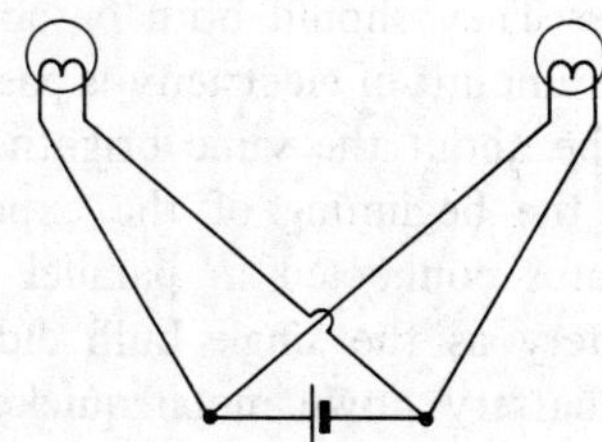

Now have four in parallel thus:

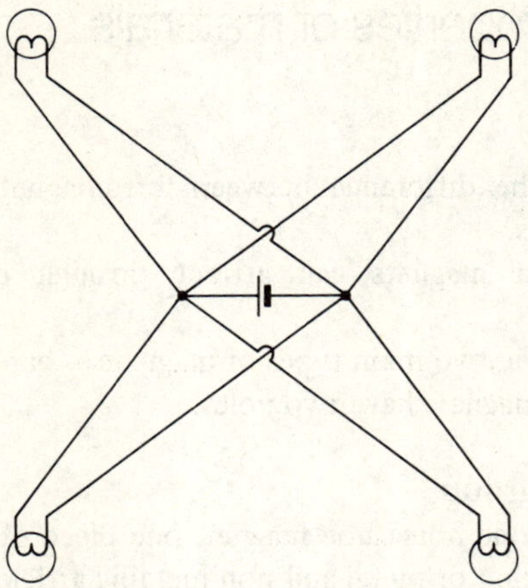

We now have the beginnings of the car lighting system. All you do is go on adding lights in parallel thus:

Beginnings of a Car Lighting System

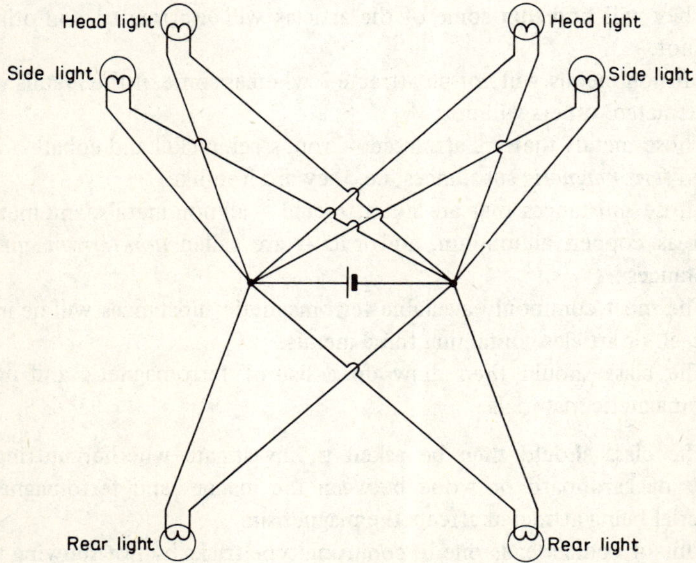

Experiment 10
Magnetic properties of materials

Objectives
a To establish the difference between ferromagnetic and non-ferromagnetic materials.
b To show that magnets can attract through non-ferromagnetic materials.
c To introduce the two main types of magnets — bar and horseshoe.
d To show that magnets have two poles.

Apparatus per group
One bar magnet, one horseshoe magnet, one piece of nickel, one piece of cobalt. A selection of metal and non-metallic articles.

Procedure
i Issue one bar magnet to each group, and ask them to bring their magnets close to any articles they can find or those that you have provided.

They will find that some of the articles will be attracted, and others will not.

All non-metals will *not* be attracted, whereas some of the metals will be attracted, others will not.

Those metals that are attracted — iron, steel, nickel and cobalt — are called *ferromagnetic* substances, i.e. they are iron-like.

Those substances that are not attracted — all non-metals, and metals such as copper, aluminium, and lead — are called *non-ferromagnetic* substances.

The most commonly available ferromagnetic substances will be iron or steel, or articles containing these metals.

The class should then draw up a list of ferromagnetic and non-ferromagnetic materials.

ii The class should then be asked to investigate whether putting a piece of cardboard or wood between the magnet and ferromagnetic material being attracted affects the magnetism.

This of course leads one to conjuring type tricks by not showing the magnet initially but asking what is making a paper-clip move across a cardboard surface without anyone apparently touching it.

iii The two previous experiments should then be repeated using a horseshoe magnet. From the results the class should be able to conclude that a horseshoe magnet is essentially a bar magnet bent into the shape of a horseshoe.

iv Give out six or so paper clips or drawing pins to each group and let them pick them all up using one end of the magnet. Ask them if they notice anything about the position of these articles.

They should notice that the paper clips or drawing pins gather around the ends of the magnet. This can then be repeated using the other end of the magnet, and will hence establish that a magnet has *two* points from which the attraction occurs. These points are called the *poles* of the magnet.

Notes
1 This last experiment can be repeated using iron filings except that they are difficult to clean off the magnets.

Experiment 11
Properties of the poles of a magnet

Objectives
a To establish that *like* poles *repel*, and *unlike* poles *attract*.
b To show that the test for a magnet is repulsion.
c To investigate the strength of a magnet.

Apparatus per group
Two magnets, one piece of unmagnetised ferromagnetic material e.g. a nail, a selection of paper clips or drawing pins.

Procedure

i Give out two magnets to each group and ask the class to investigate what happens when the magnets are brought close together.

They should thus establish that either *repulsion* or *attraction* occurs depending upon which way round they have the magnets.

On any given magnet one end is usually marked with a coloured spot (usually red) or labelled N, and thus some of the children may well see that repulsion occurs between the two marked ends or between the two unmarked ends, i.e. *like poles repel.*

If a marked end is brought close to an unmarked end of the magnet attraction occurs, i.e. *unlike poles attract.*

ii For this part of the experiment the children require only one magnet and the unmagnetised piece of ferromagnetic material.

Suggest to the class that they place the unmagnetised material on the table and bring the magnet near to it. Repeat the procedure using the other pole of the magnet. Then suggest that they place the magnet on the table and bring the unmagnetised material near to it. Repeat the procedure using the other end of the unmagnetised material.

In all cases there will have been attraction and the only object free to move was that on the table. The question should then be asked: "If you did not know which was the magnet, would this experiment have helped you to decide, since in all cases you obtained attraction and the only object to move was that on the table?"

From the resulting discussion it should be possible to decide that attraction is not a reasonable test to discover which is the magnet.

The only reasonable way is to include a second magnet and show that between magnets you can get attraction and repulsion, whereas between magnets and unmagnetised material you only get attraction. Thus the only true test for magnetism is *repulsion.*

iii The class should now be issued with a bar magnet and a fair number of paper clips or drawing pins.

The strength of the magnet can then be estimated by finding the weight it will just support, and this can be discovered by finding the number of paper clips or drawing pins the magnet can pick up as shown in the diagram.

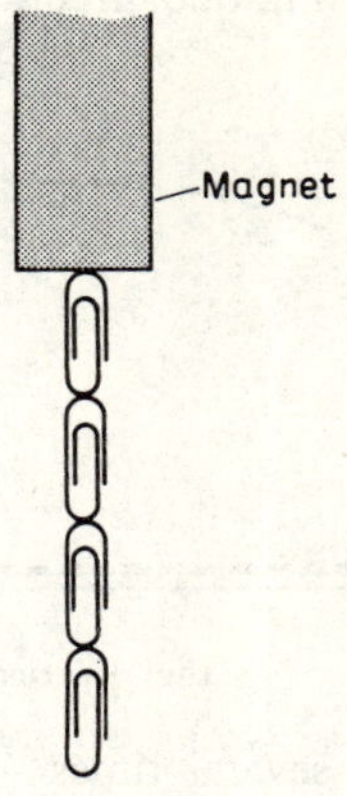

Notes

1 The poles of the magnet are called North and South poles – see Experiment 13 for the reason.

2 In Experiment (iii) care should be taken to see that a chain of paper clips or drawing pins with only one touching the magnet is produced.

Experiment 12
Making magnets

Objectives

a To make a magnet from a steel object by rubbing with a permanent magnet.

Apparatus per group

One bar magnet, one steel knitting needle.

Procedure

i It is advisable to demonstrate the technique to the whole class before allowing them to make their own magnets.

Take the knitting needle and place it horizontally. Take the magnet and stroke the knitting needle with it so that the whole length of the knitting needle is stroked in one direction, as shown in the diagram.

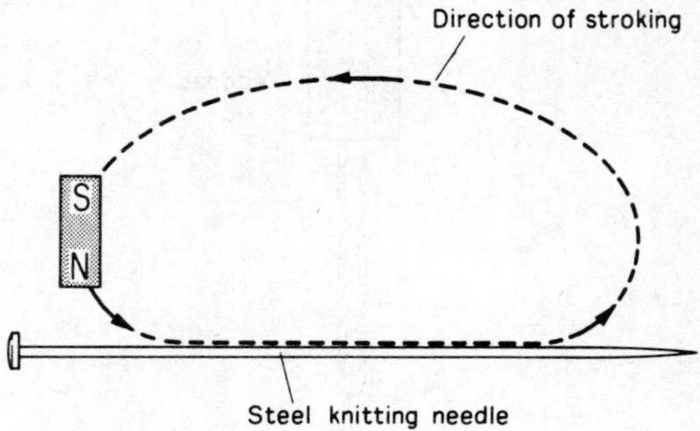

Repeat the stroking several times — the more that you stroke the stronger will be the magnetism of the knitting needle.

ii Two groups should then show by testing their two knitting needles — by repulsion — that they have in fact produced magnets.

Notes

1 When stroking the magnet always stroke in the same direction with the same pole and always lift the magnet well clear at the end of each stroke. The diagram below shows the position of the magnet at the various stages of a single stroking.

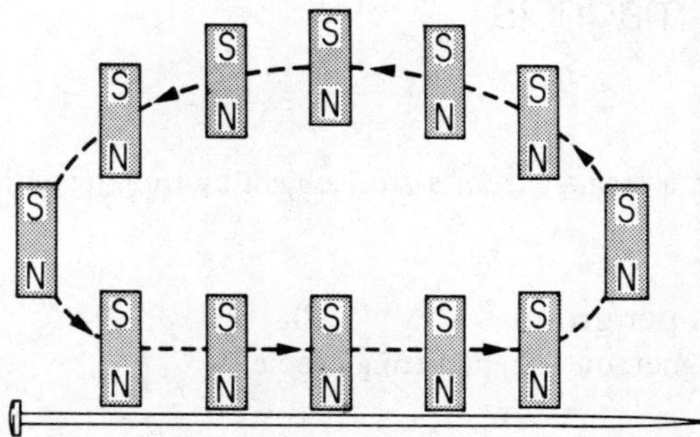

2 You do not appreciably use up your magnet by making other magnets with it, but you cannot produce a stronger magnet than you started with.

3 Your original magnet will invariably be stronger than the magnet you have been making. Hence trying to test that the steel knitting needle is

a magnet by trying to get repulsion between the original magnet and the knitting needle will probably not work, since the original magnet will almost certainly swamp the magnetism of the knitting needle, and you will always get attraction between them. Consequently it is advisable to produce two magnetised knitting needles and test for repulsion between these two.

Experiment 13
Making a compass

Objectives
a To show the main features of a compass.
b To use a small plotting compass.

Apparatus per group
One bar magnet, one magnet holder, a paper sling (already used in Experiment 3 but see apparatus construction, page 59), one steel knitting needle, a cork, one small plotting compass.

Procedure
i Give out one bar magnet and magnet holder to each group. Suggest to them that they should suspend the magnet in the holder as indicated in the diagram and allow it to come to rest. After it has come to rest suggest to them that they should gently set it swinging – as indicated in the diagram – and again let it come to rest. This they should repeat several times to see if they realise that it always comes to rest pointing in the same direction.

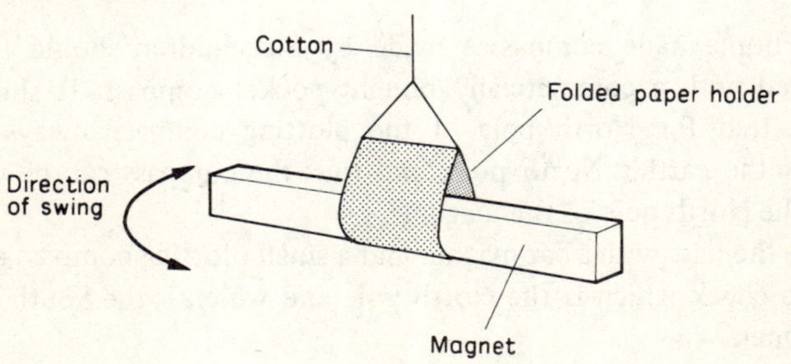

The children have produced a very crude compass, since essentially a compass is a magnet freely suspended and allowed to set in the magnetic field of the earth. The magnet will thus set North — South, with the North pole of the magnet pointing towards the earth's North pole — see Note 1 at the end of this experiment.

ii A second compass can be made using the steel knitting needle magnetised as described in Experiment 12. This can be mounted in the paper holder in the same way as the bar magnet, but will not produce a more efficient compass.

A better method is to allow the children to place the magnetised steel knitting needle upon a small cork floating in water as indicated in the diagram.

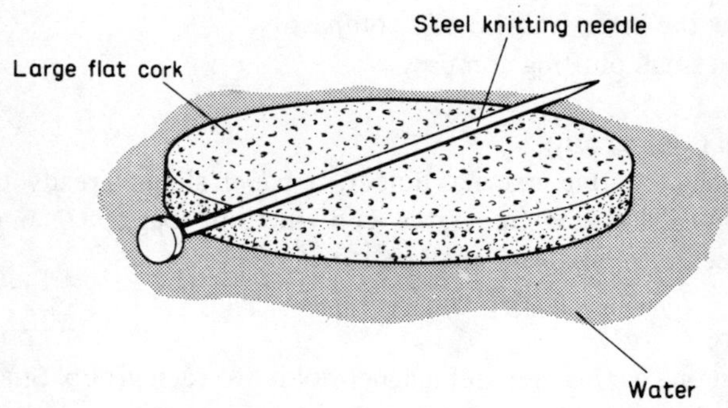

If the water is contained in a fairly shallow, flat-bottomed glass vessel it should be possible for the children to construct a paper scale marked with the points of the compass and place it under the glass container. It is set correctly by adjusting the North on the scale to coincide with the direction indicated by the steel knitting needle.

iii The home-made compasses made by the children should then be compared with a commercially bought pocket compass. It should be noticed that the North pole of the plotting compass always points towards the earth's North pole, and thus the compass can be used to check the North pole of the magnet.

Issue the class with a bar magnet and a small plotting compass and ask them to check which is the North pole and which is the South pole of the magnet.

If the plotting compass is brought to the poles of the magnet, the plotting compass will always point *away* from the North pole and *towards* the South pole as indicated in the diagram — see Notes at end of this experiment.

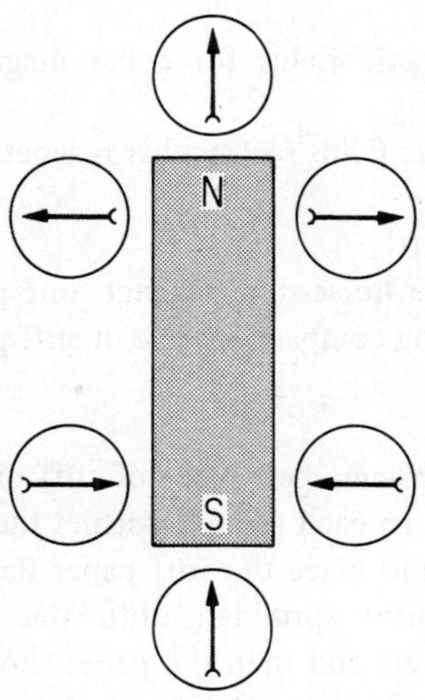

Notes

1 We should really call the North pole of a magnet, the *North-Seeking Pole,* since it seeks the earth's magnetic North pole. It is now conventional to omit the word "seeking" and merely call it the North pole. The same applies to the South pole.

2 Keep the other magnets away from the home-made or plotting compass when you are trying to locate the direction of the earth's North pole, since other magnets near at hand will affect the direction indicated by the compass.

3 Usually the pointed end of the plotting compass is the North pole of the compass.

4 Since the pointed end of the plotting compass is the North pole, and *like* poles *repel,* the plotting compass will point *away* from the north pole of a magnet and thus point *towards* the south pole of the magnet.

Experiment 14
Plotting magnetic fields

Objectives
a To plot the magnetic fields for a bar magnet and a horse-shoe magnet.
b To plot the magnetic fields for two bar magnets placed end-to-end.

Apparatus per group
Two bar magnets, one horse-shoe magnet, one pepper pot containing iron filings, one plotting compass, a piece of stiff paper.

Procedure
i Give out one bar magnet, one piece of stiff paper and a pepper pot containing iron filings to each group. Instruct the class to place the bar magnet on the table and place the stiff paper flat upon it. A few iron filings should be gently sprinkled onto the paper in the region surrounding the magnet, and then the paper should be *gently* tapped.

The iron filings should then set into a pattern similar to that in the diagram.

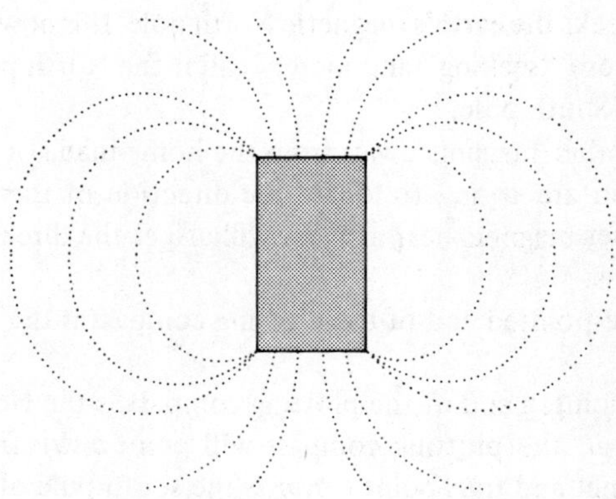

ii These lines can be investigated further by following one of them with a plotting compass as indicated in the diagram.

The children should be asked to select one of the lines and keep on placing the compass at different positions on the line. They should be asked to record what they see.

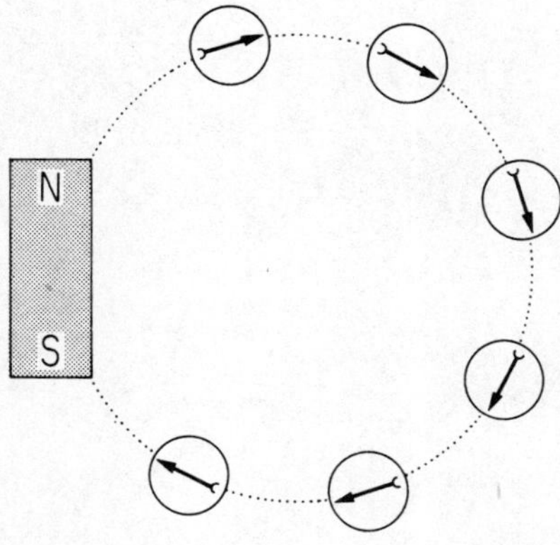

At each position the compass is trying to set along the path indicated by the iron filings. This means that we could place an arrow along this line in the direction determined by the compass i.e. going from North to South thus:

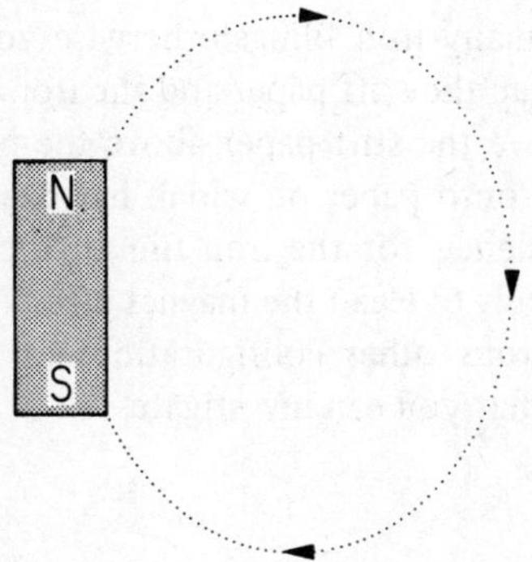

iii The procedure can now be repeated using a horse shoe magnet.

In both the cases for the bar magnet and the horseshoe magnet, the patterns can be made permanent by spraying the paper and iron filings with a clear lacquer.

iv The procedure can be repeated using two bar magnets in the end-to-end positions thus:

(a)   (b)

The configuration (b) produces an interesting case since there will be a region between the two North poles in which the lines from one magnet cancel out the lines from the other, producing a magnetic field-free region.

Notes

1 Do not use too many iron filings otherwise you will not get a very clear picture. Only tap the stiff paper and the iron filings *gently*.

2 If you do not have the stiff paper above the magnet but scatter the iron filings directly onto paper on which has been placed the magnet, there will be a tendency for the iron filings to cluster on the magnet making it very difficult to clean the magnet afterwards.

3 There are numerous other configurations of the bar magnets and horseshoe magnets that you can investigate.

Experiment 15
To show that an electric current produces magnetism

Objectives
a To show that an electric current passing through a straight piece of wire produces a magnetic field.
b To show that a coiled piece of wire acts similarly to a bar magnet.

Apparatus per group
A 2 metre length of wire, one battery, one small plotting compass.

Procedure
i Give out to each group, the wire, the battery and the plotting compass. Ask them to place the compass upon the table — it will thus be pointing N-S — and then hold the length of wire directly above the compass so that the wire is pointing in the same direction as the compass, as in the diagram.

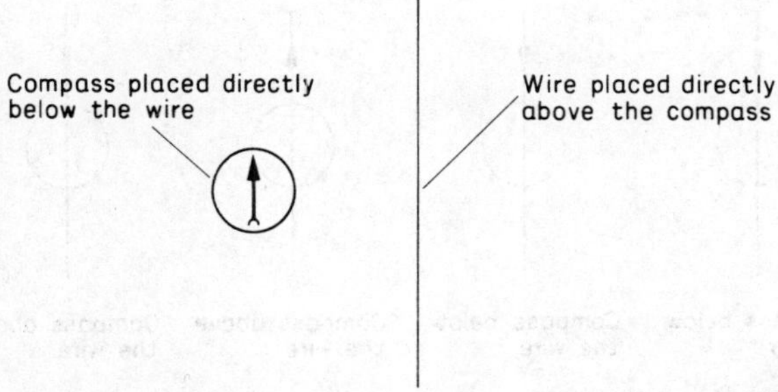

Then ask them to connect the two ends of the wire to the battery to make this circuit, making sure that the wire is still directly above the compass.

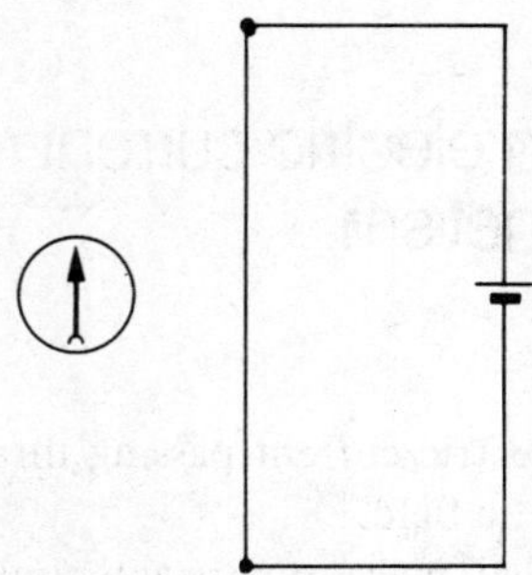

Ask the class to record anything that has happened to the compass needle — it should have been deflected either to the right or left, thus showing that the electric current passing through the wire has produced a magnetic field.

They should then be asked to perform the following experiments, each time recording what they observe:

a Reverse the current flow through the wire by reversing the connections to the battery.

b Hold the compass directly above the wire.

c Reverse the current flow through the wire when the compass is held above the wire.

Consequently they should investigate the four situations indicated in the diagram.

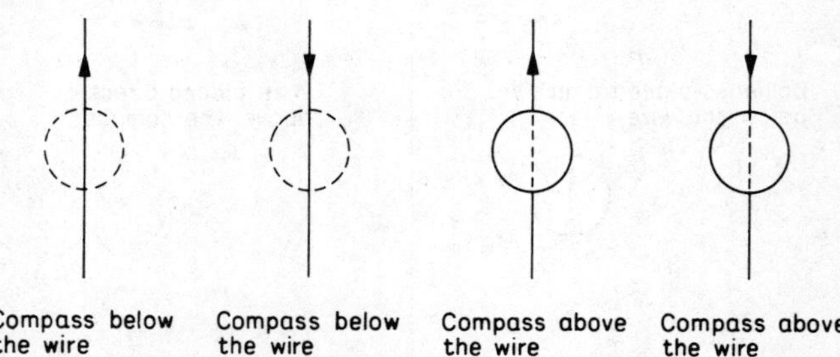

ii The class should then be instructed to make a *solenoid* — this is essentially a long coil of wire — by winding the long piece of wire around a pencil and then removing the pencil. They will thus obtain this:

They should then connect the two free ends of the wire to the battery to make the circuit.

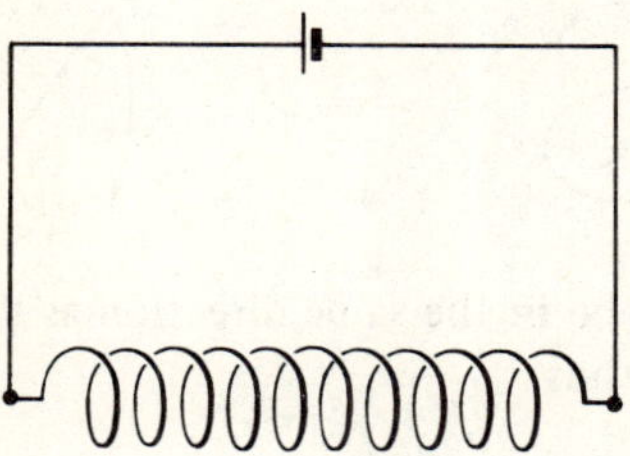

Ask them to bring the small plotting compass near to the two ends of the solenoid while the current is flowing. They should discover that one end is a North pole and the other is a South pole *(see Experiment 13 to check which pole is which)*.

Ask the class to reverse the direction of current flow through the solenoid, by reversing the connections to the battery and note if any change occurs to the compass. They should in fact find that the poles of the solenoid have changed over.

It is possible to forecast which pole will be which by considering the direction of flow through the ends of the solenoid, remembering that the electric current is said to flow from positive to negative *(see Experiment 4)*.

Look at the ends of the solenoid as indicated in the diagram.

Record the direction of current flow in the ends of the solenoid:

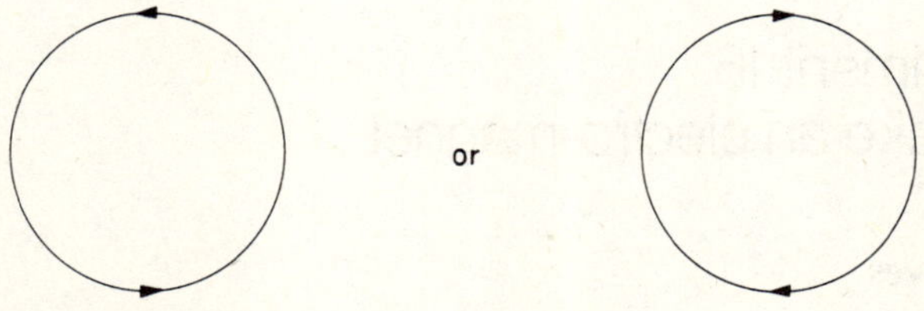

depending upon which end you are viewing from.

To forecast which end is which, you put arrows on the ends of the N and S thus:

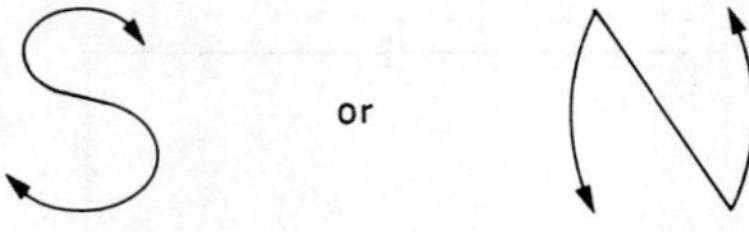

These arrows will be in the same direction as the current flow at the ends of the solenoid thus:

Consequently the children should be able to check that this is in fact true.

Notes
1 The 2 metre length of wire should be plastic-coated and have crocodile clips at both ends (see apparatus construction section, page 59).
2 The battery should be kept as far away from the plotting compass as possible.
3 It is essential to start with the wire and compass pointing in the same direction before any current is passed through the wire.

Experiment 16
To make an electro-magnet

Objectives
a To make an electro-magnet.
b To consider the uses of an electro-magnet.

Apparatus per group
A 2 metre length of wire, one battery, one 6 inch nail, a selection of paper clips, sellotape.

Procedure
i The children should repeat the second part of Experiment 15 and make a solenoid of wire. They should then show that despite it appearing to behave like a magnet in so far as it has a N and a S pole, it will not act like a magnet and pick up articles. Consequently it is necessary to modify the solenoid and this is done using a suitable metal *core*.

ii The class should be issued with 6 inch nails — these must be iron or steel since the core must be ferromagnetic. The children should then wind the coil onto the nail. It will probably be necessary to hold it in position with sellotape to prevent it unwinding.

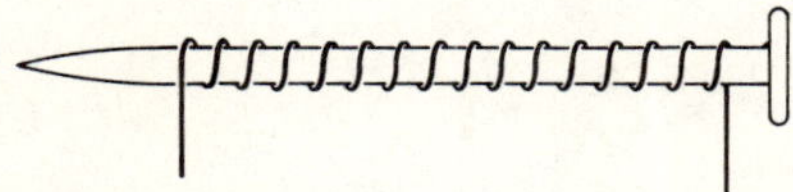

The two free ends of the wire — which should have crocodile clips attached — should be connected to the battery and the children should check that they have produced a magnet. The poles of this magnet can be forecast from the direction of the current flow as indicated in Experiment 15.

iii The children should check the strength of their magnets by seeing how many paper clips it will hold (see Experiment 11).

The children should connect two batteries in *series* (see Experiment 9) thus:

and then connect these two batteries across the electro-magnet as below:

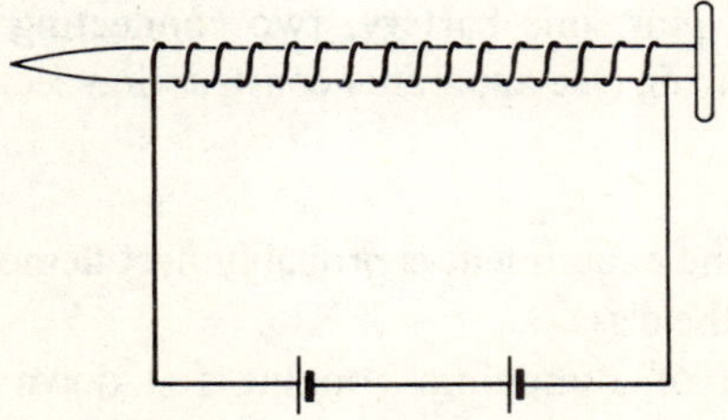

and then find out if this has increased the strength of the magnet.

They should also investigate what happens if they switch the current off.

iv There are many applications of the electro-magnet, namely in scrap yards, heavy industry and the electric bell (see additional information, page 72).

Notes

1 The 6 inch nails should be checked to be non-magnets before being used as the core of the electro-magnet. A compass can be used to check this.

Experiment 17
To show that an electric current
can produce movement

Objectives

a To show that when an electric current passes through a metal which is placed in a magnetic field, a force, and hence movement, is produced.

Apparatus per group

A 1 metre long piece of aluminium cooking foil approximately 1 cm. wide, two drawing pins, one battery, two connecting wires, two bar magnets, one wire swing (see apparatus construction section, page 63).

Procedure

i The first part of the experiment is probably best demonstrated by the teacher in front of the class.

Fasten the strip of aluminium cooking foil down onto the table reasonably slackly, using the two drawing pins. Place the two bar magnets, one either side of the foil as in the diagram.

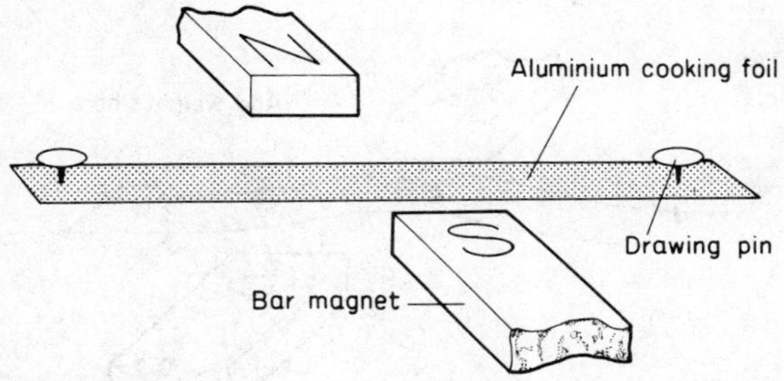

The aluminium foil is now between the poles of the magnet and is thus in a magnetic field. If a current is passed through the foil, the current will thus be passing through a magnetic field.

Now connect up the following circuit.

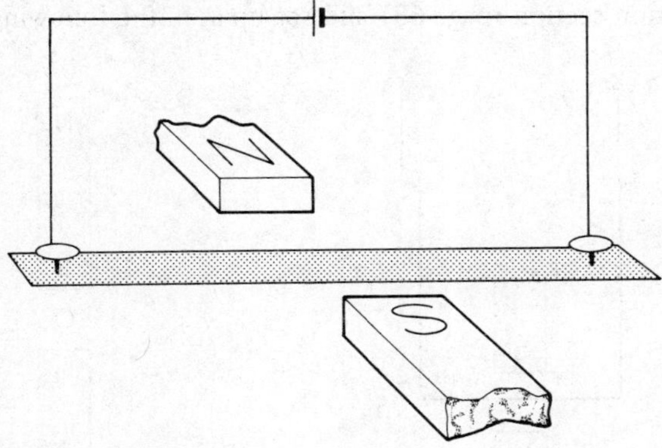

The aluminium foil will now be seen to move, showing that a force is being exerted upon it. If the battery is disconnected, the force will disappear and the aluminium will assume its original position.

ii Issue the apparatus to the class so that they can repeat the experiment. Suggest to them that they should investigate the effect of reversing the direction of the current through the aluminium foil and then reversing the poles of the magnets. In this way they should establish that the direction of the force depends upon the direction of the electric current and the direction of the magnetic field.

iii The class can investigate the strength of the force produced by adding weights to the aluminium foil until it touches the table again — as in the diagram.

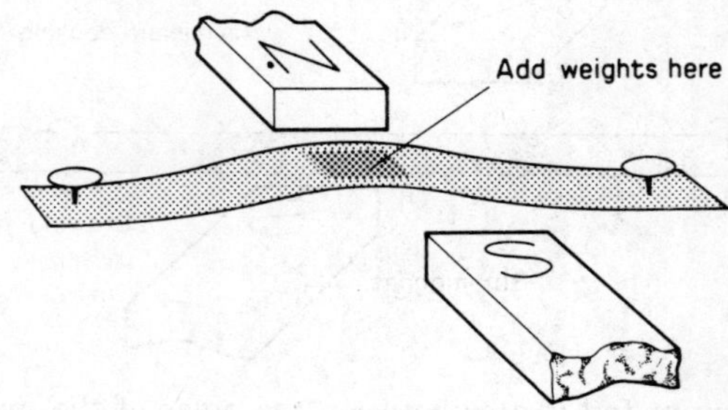

iv The next experiment is probably best demonstrated by the teacher in front of the class.

The wire swing should be constructed as indicated in the apparatus construction section (page 63) and set up as in the following diagrams:

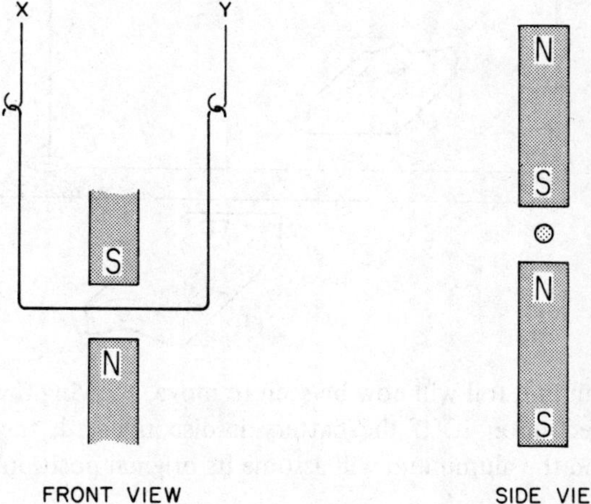

The two magnets must be placed above and below the swing as indicated in the diagram. The battery should be connected between the points X and Y, thus an electric current will be flowing through the wire in the magnetic field, and again a force is produced which moves the wire swing out of the magnetic field.

Check that the wire swings back again if the battery is disconnected.

The direction of the force, i.e. the movement of the swing, can be investigated by reversing the direction of the electric current flow and reversing the magnets.

Notes

1 Make sure that the poles of the two bar magnets are as in the diagram, and always have a N opposite to ₂ S, and not two N's opposite each other.

2 In the swing experiment make sure that the two bar magnets are positioned as shown in the side view.

3 The dry battery may not produce sufficient electricity to show an appreciable movement, and hence a stronger battery — having a voltage greater than 3 volts — should be used. A car battery would be ideal for this.

Experiment 18
To construct a simple electric motor

Objectives

a   To construct a simple electric motor.

Apparatus per group

One cork 2 cm. in diameter, 4 cm. long, a thin aluminium knitting needle about 20 cm. long, three pieces of soft wood one 12 x 8 x 1 cm., and two 8 x 8 x ½ cm., two pins or thin nails approx. 4 cm. long. Thin insulated copper wire for the coil. Thick copper wire about 20 cm. long, two cup hooks, battery, bulb, one horse-shoe magnet or two bar magnets, as strong as possible.

Procedure

i   The details that follow are a stage-by-stage series of instructions regarding the building of the simple electric motor.

ii We have seen that if an electric current passes through a magnetic field a force and hence movement is produced. Consequently the first task is to prepare the wire — in the form of a coil — that will carry the electric current through the magnetic field. The coil can be wound upon a cork, the cork will rotate in the magnetic field and hence must be mounted upon an axle. A suitable axle is a thin aluminium knitting needle. The axle needs to be non-ferromagnetic and as light and strong as possible and therefore must *not* be made from steel. Push the needle through the middle of the cork from end to end taking care that it goes exactly through the middle as in the diagram.

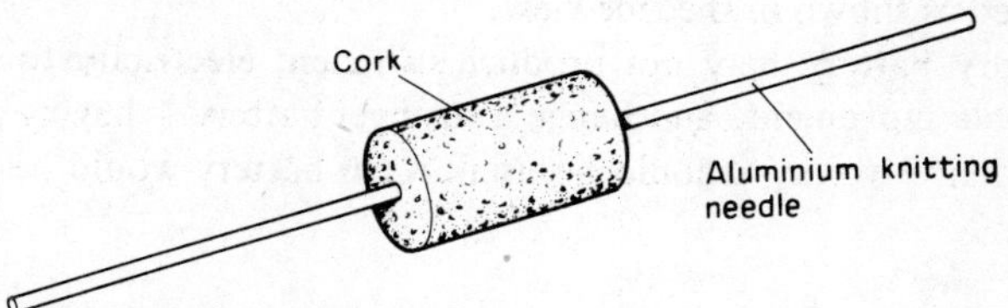

iii Using a very sharp knife or razor blade the cork can now be cut into a rectangular shape thus:

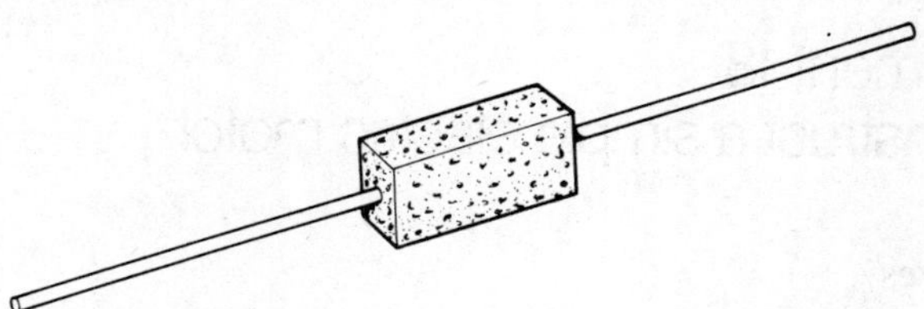

and the two top edges either cut or filed thus:

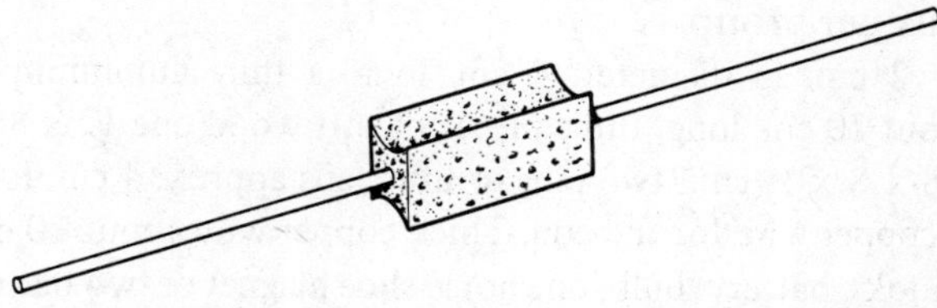

iv The axle and cork have to rotate in a magnetic field and hence must be mounted to do so. Using the three pieces of wood, a frame should be constructed as indicated in the diagram. Two holes should be drilled in the two end pieces as indicated so that the aluminium axle can spin freely in them.

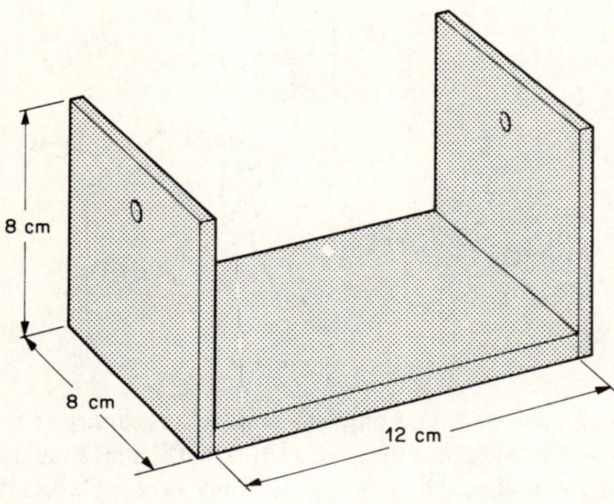

v The arrangement of the axle, cork and framework is as in the following diagram. Make sure that the coil can spin well clear of the base-board.

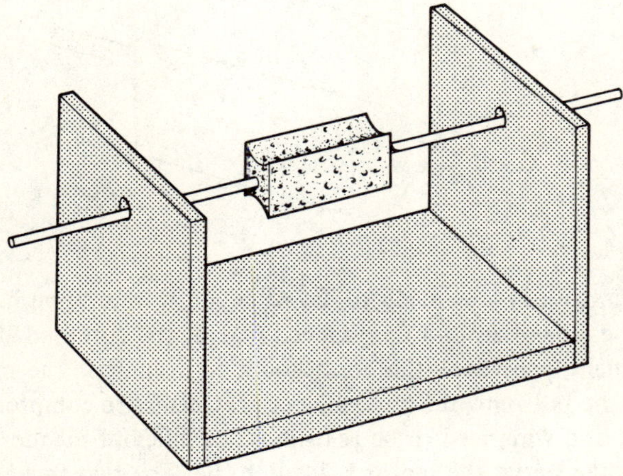

vi The cork and the axle will tend to slip sideways and so to prevent this two pieces of cork may be used as spacers. The next diagram indicates how the spacer is fitted on one side, the other side being fitted in a similar manner.

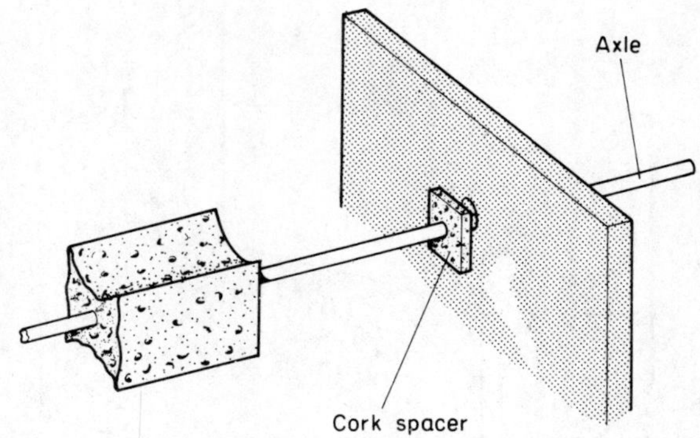

vii Before the coil is wound onto the cork it is necessary to fasten two long thin pins or nails into the cork as shown. The pins should stick out for at least 2 cm., should be driven in as firmly as possible and should be as near to the outside of the cork as possible, so that the distance "x" equals exactly the distance "y".

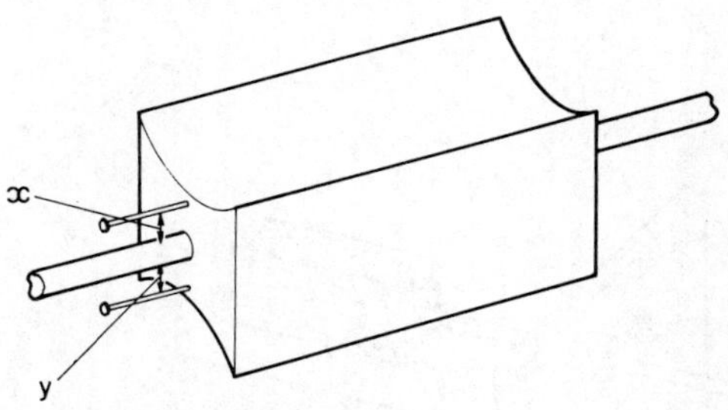

viii   The cork is now ready to have the coil wound upon it. The wire should be wound around the grooved part of the cork and all the wire that actually goes onto the cork should be insulated. The more turns that can be put onto the cork the better. One has to compromise since the cork and wire need to be as light as possible, but the more turns of the wire the better the motor is likely to be. The two free ends of the wire should be bared and one attached firmly to each nail. The bared ends of the wire should be wound around the nail and should be kept as near as possible to the cork, but should not touch the bare end connected to the other pin or the axle. The coil should appear as in the diagram.

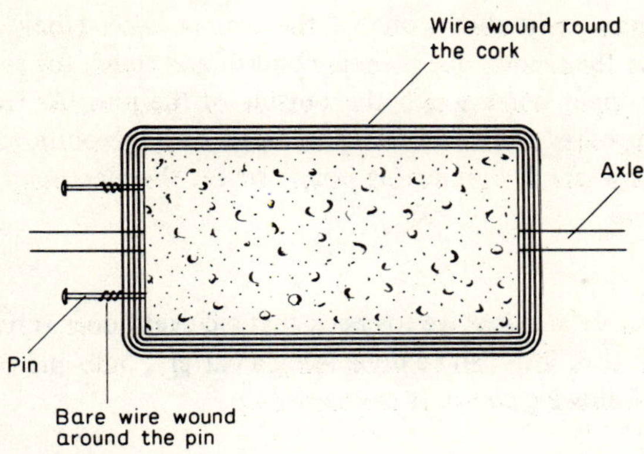

ix Since the coil will be rotating, it is impossible to use ordinary connecting wires to connect the battery to the coil, as these would soon become tangled up, and so the two pins are used to get the electricity into and out of the coil.

x Two thick pieces of copper wire must now be arranged vertically so that as the coil rotates the two pins on the coil will brush against these two pieces of copper wire. This can be arranged by screwing two cup hooks into the base board and fastening one piece of copper wire under each cup hook. The copper wire is then bent upwards until it touches the pin in the coil as indicated in the diagram.

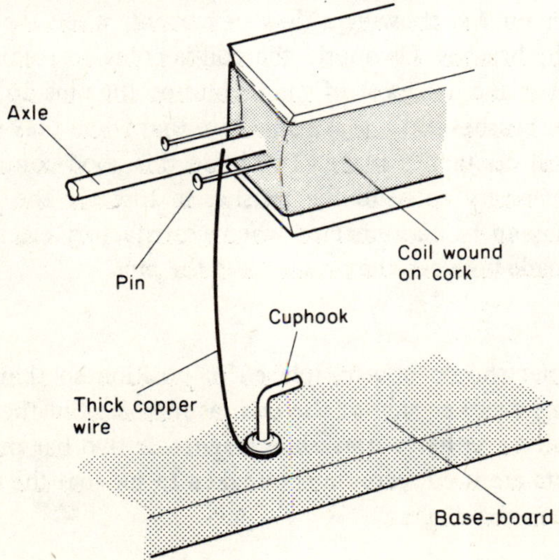

The diagram only shows one of the copper wires (they are called brushes), but the second one is similar and should touch the second pin. The thick copper wires touch the outside of the pins. As the coil rotates these brushes will touch the pins once every revolution. The two cuphooks now act as connecting points to get the electricity into and out of the coil.

xi It is advisable at this stage to check that a current does actually flow through the coil. This can be done using a battery, bulb and connecting wire, if the following circuit is connected up.

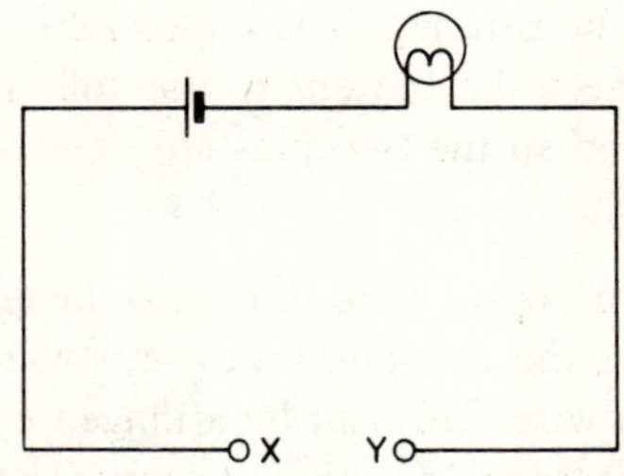

The two points X and Y are the two cuphooks. It should be found that the light is on, i.e. showing a flow of current, when the two pins are touching the brushes. Obviously the coil is going to rotate when it works as a motor and for most of the revolution the pins do not make contact with the brushes and so it is imperative that when they do touch, a good electrical contact is made. To obtain this good contact it will probably be necessary to bend the brushes in towards the pins. The above circuit should be disconnected when a satisfactory electrical contact has been made between the brushes and the pins.

xii   The magnet should now be placed in position so that the two poles are arranged either side of the coil, as indicated in the diagram. The magnet can be either a horseshoe magnet or two bar magnets. If two bar magnets are used, care must be taken to see that the two poles are as indicated in the diagram.

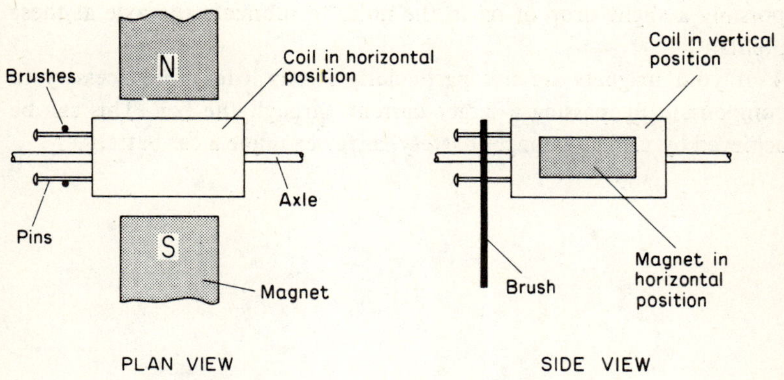

xiii   Having placed the magnets in position the electrical circuit indicated below should be connected up. It is usually necessary to have a fair number of batteries in the circuit since a fairly large current is required to drive the motor. X and Y are, as before, the two cuphook connectors.

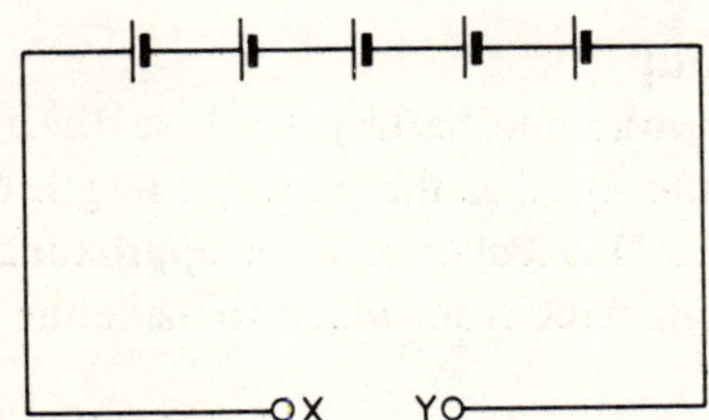

xiv   To get the motor to work it may still be necessary to adjust the brushes and to give the coil a spin.

Notes

1 If the motor still does not work even though you are quite happy about the electrical contacts, it may well be that the magnet is not strong enough, since the stronger the magnet the better the motor works.
2 The more turns of wire the better the motor should work, but care must be taken to see that the coil is not made too heavy.
3 Make sure that the coil on the axle is as free to spin as possible. This may mean having reasonably large holes in the end pieces of wood, and

possibly a slight drop of oil in the holes to lubricate the axle at these points.

4 If your magnets are not particularly strong it may be necessary to compensate by passing a larger current through the coil. This can be achieved by using a stronger battery, as for example a car battery.

Experiment 19
Experiments with electric motors

Objectives
a To perform simple tasks with small toy electric motors.

Apparatus per group
One toy electric motor, one battery to drive the motor, one pulley wheel to attach to the shaft of the motor, a length of twine or string, a selection of weights. One length of wood approximately 1 metre long, 1 cm. thick and 10 cm. wide from which to make the inclined plane.

Procedure
i In performing this series of experiments it is advisable to buy a small toy electric motor, since they can be bought for a few pence. Alternatively, if the electric motor you have made is good enough this could be used. It is also possible, by performing these experiments with a bought motor and the home made one, to compare the efficiency of the one you have made with the bought one. The details and discussions that follow are presupposing that these experiments will be demonstrated by the teacher, but if sufficient apparatus exists the children can easily do them themselves.

ii Attach a large pulley wheel — one from a meccano set — onto the axle of the motor as in the diagram.

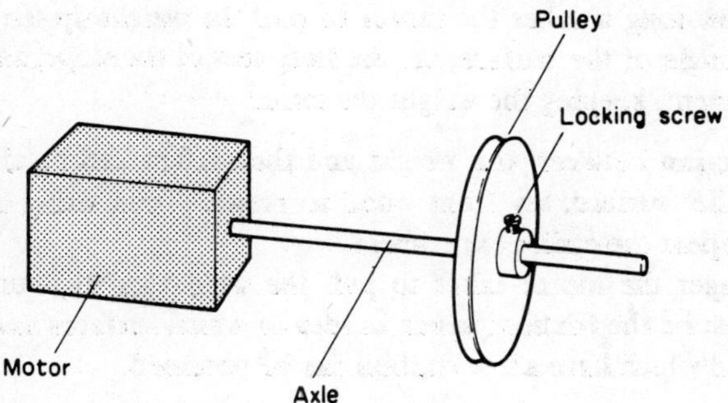

Fasten a long piece of string to the rim of the pulley wheel and place the motor and battery at the edge of a table as in the diagram.

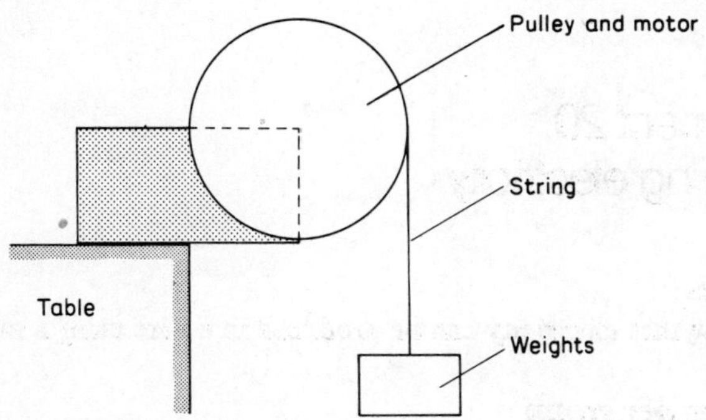

Attach various weights to the end of the string and time how long it takes the motor to lift them through a certain distance. Find the largest weight that the motor can just lift.

iii Make a sloping surface as indicated in the diagram.

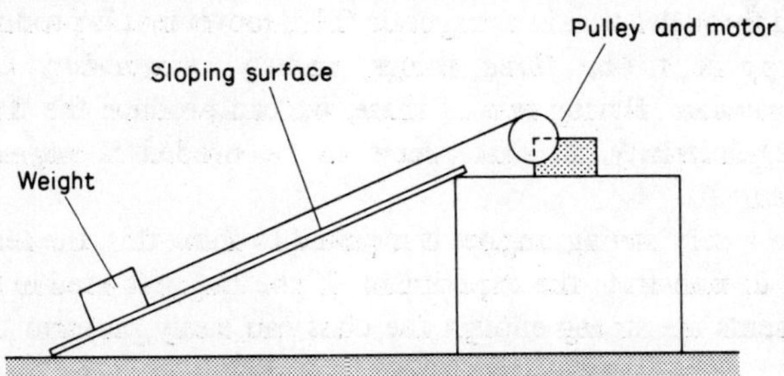

Find how long it takes the motor to pull the weight up the surface. Alter the angle of the surface, i.e. the steepness of the slope, and repeat the experiment, keeping the weight the same.

iv The friction between the weight and the surface can be altered by changing the surface, say from wood to plastic. Thus, using different surfaces, repeat the earlier experiment.

The longer the motor takes to pull the weight up the surface the greater must be the friction; hence an idea of which surfaces have a high friction and which have a low friction can be obtained.

Experiment 20
Producing electricity

Objectives
a To show that electricity can be produced in a wire using a magnet.

Apparatus per group
One plotting compass, one strong magnet, two lengths of wire, each approximately 3 metres long.

Procedure
i In the previous experiments we have seen that if an electric current is passed through a wire in a magnetic field, movement is produced. Thus there appear to be *three* things, namely, *magnetism, electricity,* and *movement.* Having two of these, we can produce the third. So to produce electricity all that seems to be needed is *magnetism* and *movement.*

Since a very strong magnet is required to show this, the teacher may have to demonstrate the experiment. If the magnets used in the earlier experiments are strong enough the class can easily perform the experiment.

ii Wind one of the lengths of wire into a coil as in the diagram:

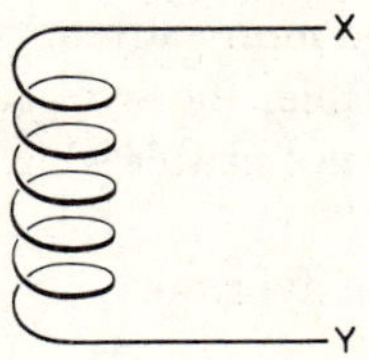

Connect the two free ends XY, to the ends of the second long piece of wire as in the diagram.

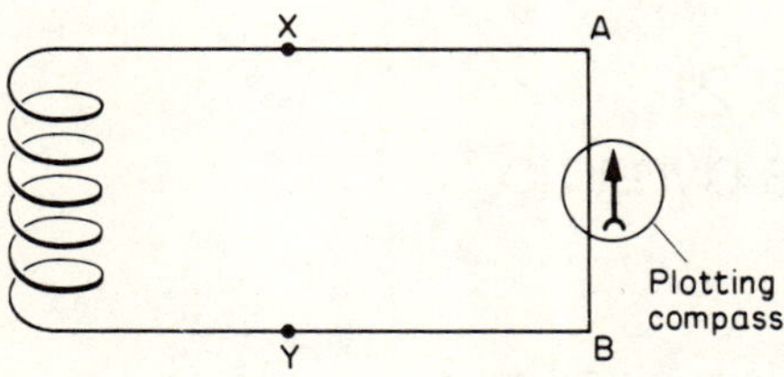

The second piece of connecting wire should be bent into the shape XABY so that the length AB is about 30 cm. and the plotting compass, placed under the wire, is well away from the coil. The wire AB should be arranged to be pointing in the same direction as the compass is pointing.

The compass is to be used to detect whether a current flows through the wire, since we have seen in an earlier experiment that an electric current produces a magnetic field and will therefore deflect the compass needle. The fact that the compass needle is initially undeflected means that no current is passing through the wire.

Move the magnet very quickly into and out of the coil and observe that there is a deflection of the compass, showing the production of an electric current in the wire. The deflection only occurs when the magnet is being moved into or out of the coil, therefore supporting the statement that movement is essential. The coil must be as far away from the compass as possible so that the magnet will not affect the compass as it moves in the coil. If it is thought that the magnet is affecting the compass it will be necessary to use much longer connecting wire to move the coil still further away from the compass.

Notes

1 You will need a fairly strong magnet to observe this current, since the stronger the magnet the more current will be produced. If only a horse shoe magnet is available, the wire can be moved between the poles of the magnet. It does not matter which moves, the magnet or the wire.

2 This is the principle of the dynamo.

Experiment 21
The simple dynamo

Objectives

a To show the principles behind a simple dynamo.

Apparatus per group

One bulb (1.5V or less), one bulb-holder, a bought electric motor, a pulley to attach to the motor axle, length of string, selection of weights.

Procedure

i A dynamo is essentially an electric motor except that the axle, and hence the coil, is rotated manually and consequently electricity will be produced in the coil. For this experiment it is better to use the bought electric motor, but your home-made one may work.

ii Instead of connecting a battery across the motor, a 1.5 volt. bulb is now connected in its place. With the large pulley on the axle drive the axle as fast as possible, using in the first instance your hand. You should observe the light coming on, showing that the motor is now acting as a dynamo and hence producing electricity.

The problem is how to rotate the pulley as fast as possible. One method the class should try is to wind a string around the pulley and attach the free end to a heavy weight. As the weight is released it undoes the string on the pulley and therefore turns the axle. The class could then investigate the brightness of the bulb for different weights falling through different distances.

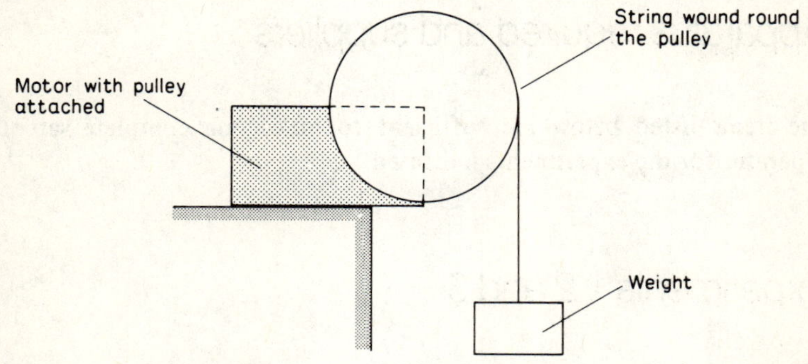

iii Another method of turning the pulley is to construct some form of water wheel, e.g. by adding vanes to the pulley and letting water from the tap drive it. Another possibility could be the construction of some form of windmill. These of course lend themselves to projects since each group could devise and build different methods of driving the dynamo.

iv If there is one available, a bicycle dynamo can be opened up and looked at. These are driven by making contact with the tyre of the cycle, since as the tyre goes round it drives the coil in the dynamo.

Notes
1 The electricity produced is called A.C. (alternating current) as opposed to D.C. (direct current) produced from a battery. This, however, is a topic that should be avoided at this stage since it is quite complicated.

Section Two

Apparatus required and suppliers

The items listed below are sufficient to make one complete set of apparatus for the experiments indicated.

Experiments 1, 2 and 3

Selection of as many different plastics as possible in strips approximately 2 x 20 cm.

Two strips of perspex, approximately 2 x 20 cm.

Two strips of cellulose acetate, approximately 2 x 20 cm.

A selection of domestic plastic appliances e.g. toothbrush, biro, and ruler.

Comb — one plastic and, if available, one metal.

Selection of metals, e.g. copper, aluminium, and iron. The size of these is not particularly important.

Tissue paper.

Two balloons.

Reel of cotton.

Drawing pins.

A wooden ruler.

A paper sling (see apparatus construction, page 59).

The strips of plastics can usually be obtained either from model toy shops or factories dealing with plastics, since they will usually have an abundance of off-cuts. The remaining articles are all easily available from stores such as Woolworths.

Experiments 4, 5, 6, 7, 8 and 9

Apparatus indicated * can be obtained cheaply from:

Radio Spares Components Ltd.,
PO Box 427,
13-17 Epworth Street,
London, EC2P 2HA.

Apparatus indicated + can be obtained cheaply from an electrical shop.
Apparatus indicated X can be obtained cheaply from a hardware shop.
Apparatus indicated ϕ can be obtained cheaply from a chemist.

+ One battery (twin cycle battery, 3 volts. Type 800).
* One bulb (2.5 volts, 0.2 amps. Round M.E.S. Pilot type).
* Two bulbs (3.5 volts, 0.15 amps. Round M.E.S. Pilot type).
* Two bulb-holders (M.E.S. Battenholder – to hold above bulbs).
* Eight crocodile clips (standard type).
* Electrical connecting wire (type 1/0.6 mm. white).
+ One Bib wire stripper.
+ One electrician's screwdriver.
X A pair of pliers.
   A reel of sellotape.
X Six ¾ inch right-angled metal dresser hooks.
   One 4 inch length of straight meccano.
   One piece of soft-wood, approximately 6 x 2 x ½ inch.
   One piece of soft-wood, approximately 3 x 2 x ½ inch.
   Two pieces of soft-wood, approximately 2 x 2 x ½ inch.
   One piece of soft-wood, approximately 4 x 2 x ½ inch.
X Two 6 inch nails.
X Two No. 4, ¾ inch screws.
X Two 1 inch thin nails.
   One small jam-jar or equivalent size glass container.
X Emery paper.
X Steel wool.
   Selection of large bottles to hold the liquids.

Selection of Solids

Rusty nails, wood, paper, combs, plastics, different types of metals, pencil sharpened at both ends, wool, string, and any others you can think of.

Selection of Liquids

a Water.

b Milk.

c Ink.

d Paraffin.

e Acid from a car battery or an accumulator.

f Vinegar.

Solids to add to water

    a Common cooking salt.

    b Washing soda.

    c Bath salts.

ø d Copper sulphate crystals.

ø e Magnesium sulphate (Epsom salts).

    f Detergents.

Experiments 10, 11, 12, 13 and 14

Apparatus indicated * can be obtained cheaply from a toy or hardware shop. Other items except the nickel and cobalt can be obtained from a store such as Woolworths. The nickel and cobalt can be obtained from a steel works or a metal company such as Mond Nickel Company, Thames House, Millbank, London, SW1.

* Two bar magnets.

* One horse-shoe magnet.

  Paper clips.

  Drawing pins.

  Selection of ferromagnetic and non-ferromagnetic materials.

  Piece of nickel and cobalt.

  Two 6 inch steel nails.

Two steel knitting needles.
A flat cork, approximately 10 cm. in diameter and 2 cm. thick.
One plotting compass.
* A pepper pot containing iron filings.
Piece of stiff paper or thin cardboard.
Clear lacquer, e.g. an aerosol of hair lacquer.
Paper sling (see apparatus construction, page 59).

Experiments 15, 16, 17, 18, 19, 20 and 21

The following items, required for the earlier experiments, which are also required for these experiments, are:
Connecting wire.
Battery (twin cycle battery, 3 volts. Type 800).
One bulb and bulb-holder.
One small plotting compass.
One 6 inch steel nail.
Paper clips.
Sellotape.
Drawing pins.
Two bar magnets, as strong as possible.
One horse shoe magnet, as strong as possible.
Two cup hooks.

The following items are also required to complete the apparatus for these experiments. They can all be obtained from an electrical, a toy, or a hardware shop.
Aluminium cooking foil.
Wire swing (see apparatus construction, page 63).
A cork, approximately 2 cm. in diameter and 4 cm. long.
A thin aluminium knitting needle about 20 cm. long.
Two pins or thin nails approximately 4 cm. long.
Very thin insulated copper wire.
A 20 cm. length of thick copper wire.
A toy electric motor and battery to drive it.

One pulley wheel capable of being attached to the drive shaft of the above toy motor.
A length of string or twine.
A selection of weights.
A piece of soft-wood, approximately 12 x 8 x 1cm.
Two pieces of soft-wood, approximately 8 x 8 x½ cm.
One piece of soft-wood approximately 1 metre long by 1 cm. thick by 10 cm. wide.

A great deal of the equipment required for these, and the later experiments is supplied commercially by the following suppliers. Price lists can be obtained directly from them.

Philip Harris,
Frederick Street,
Birmingham B1 3DJ.

Griffin & George Ltd.,
Ealing Road,
Alperton,
Wembley,
Middlesex HA0 1HJ.

Section Three

Apparatus construction

Paper Sling — for electrostatics and magnetism (for Experiments 3 and 13).

Take a piece of paper approximately 10 x 20 cm. Fold it to make two squares 10 x 10 cm. Make a hole as indicated in the diagram. Pass a piece of cotton through this hole, which will have the added effect of joining the two edges of the paper together. Suspend as indicated.

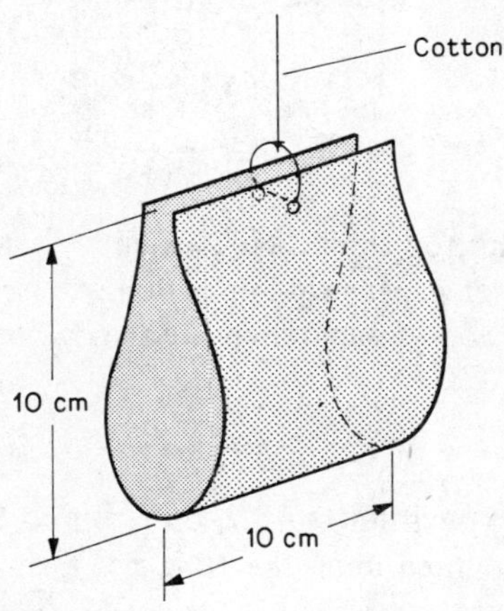

Connecting Wire

Cut three 50 cm. lengths of the wire and bare the ends, removing approximately 2 cm. of the plastic covering at each end. Unfasten the screw on the crocodile clip, push the wire through the cylindrical end, fasten the wire under the screw and tighten the screw up. Repeat so that all three pieces of wire have crocodile clips attached to the ends.

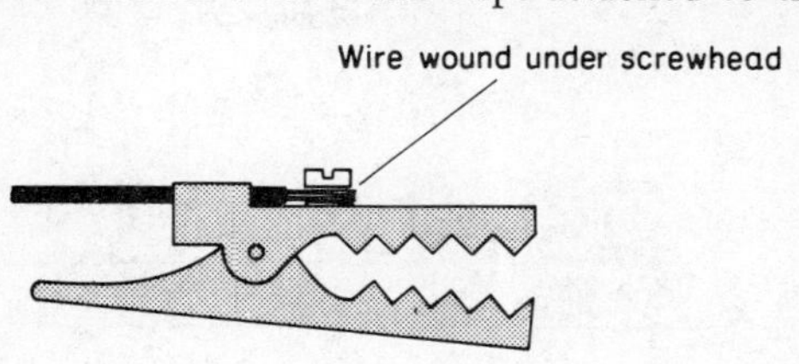

Battery

The terminals of this battery are not labelled, and so it is necessary to label them "+" and "−"

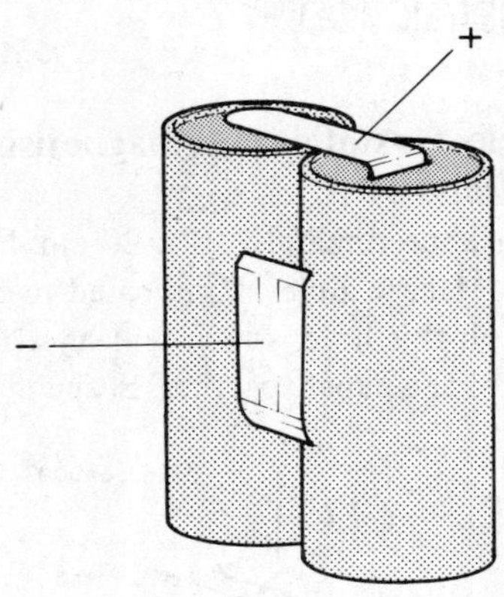

The terminal at the top should be labelled "+" and the one in the middle "−". A piece of paper held in position with sellotape should be sufficient. It is also a good thing to indicate the "+" terminal RED, and the "−" terminal BLACK.

Bulb-holder (for Experiments 4, 5, 6, 7, 8 and 9).

The bulb-holder is mounted upon the 10 x 5 x 1 cm. wood, using the two ¾ inch No. 4 screws.

Screw two of the dresser hooks into the wood as shown in the diagram. These two hooks will act as terminals and therefore they must be connected by short pieces of wire to the two terminal bolts on the bulb-holder.

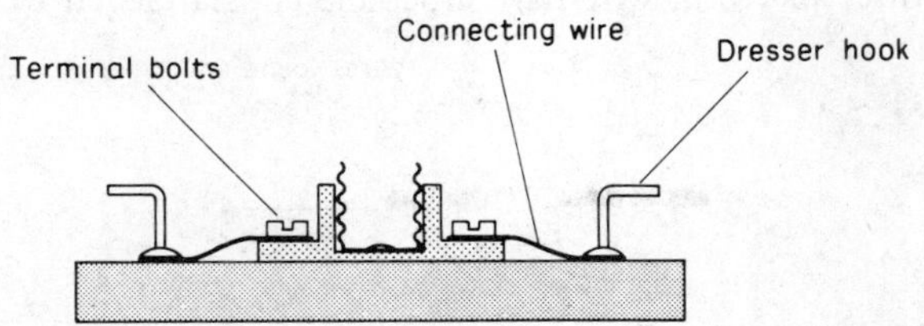

Tapping Key (for Experiment 5).

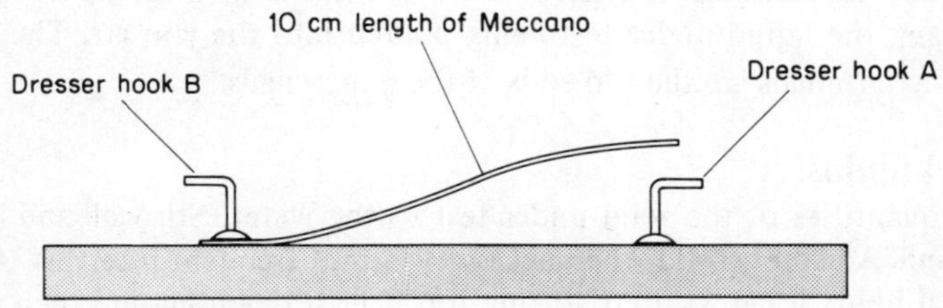

Scrape off the paint from the piece of meccano, giving a final clean with the emery paper. Screw the dresser hook A in position. Bend the piece of meccano as indicated in the diagram, and fasten to the wood — approximate size 15 x 4 x 1 cm. — by screwing the dresser hook B through one of the holes in the meccano. The meccano should not touch A unless pushed downwards, in which case on releasing the meccano it should spring back to position indicated in the diagram.

The connecting terminals are the dresser hooks A and B, and the tapping key is worked by pushing down the free end of the meccano until it touches A and then releasing it.

The Liquid-Tester or Voltameter (for Experiment 8).

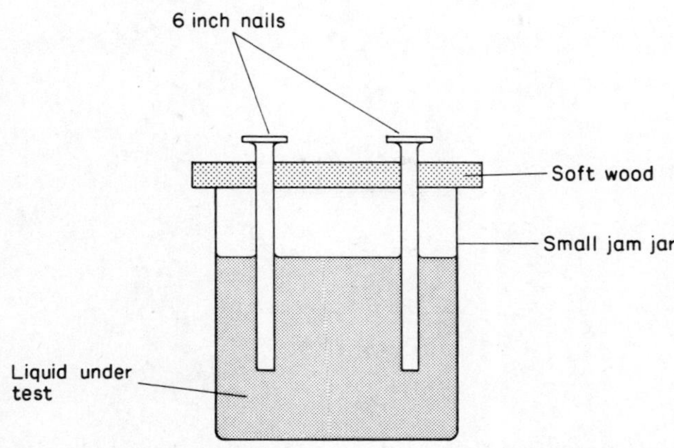

The pointed ends of the 6 inch nails should be removed, either with a saw or a file. Two holes, about 2 cm. apart are drilled in a piece of

soft-wood — approximate size 6 x 4 x 1 cm. — so that the 6 inch nail
will just slide through the hole. The apparatus is then set up as in the
diagram, the liquid under test being poured into the jam jar. The connecting terminals are the top ends of the 6 inch nails.

The Liquids

Add quantities of the solid under test to the water. Stir well and leave
to stand. Add more solid when necessary, stir at frequent intervals. After
several hours leave without stirring for at least one hour and then pour
the liquid into the bottle without disturbing the remaining solid.

The Fuse Tester (for Experiment 6).

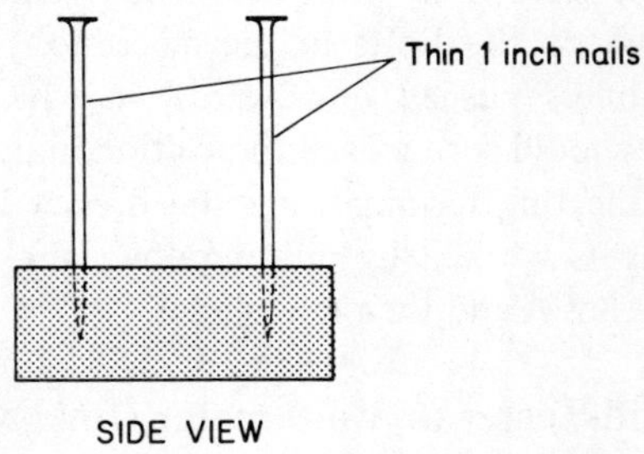

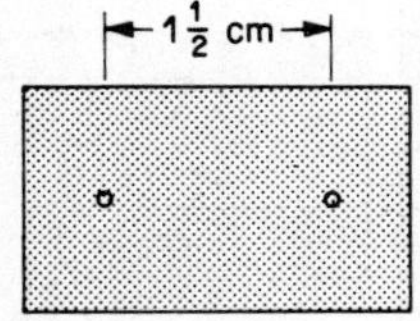

Hammer the two nails into the wood so that they are held firmly in
place, but not buried too deeply. The nails should be about 1½ cm.
apart.

The Wire Swing (for Experiment 17).

Take a piece of bare copper wire, about 20 S.W.G. (standard wire gauge) thick, and 30 cm. long. Bend it into the shape indicated in the diagram below, so that on each free end there is a hook.

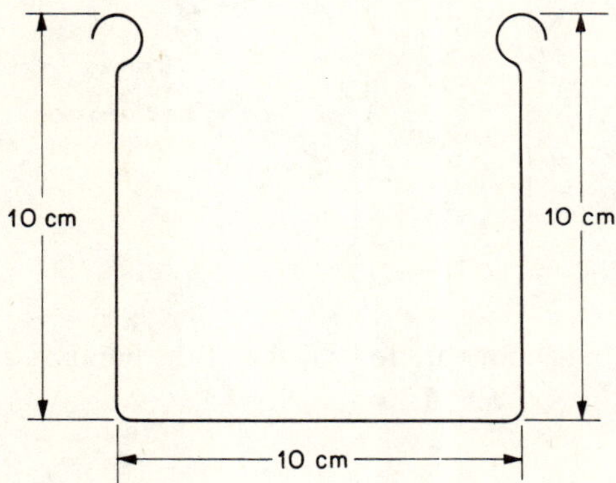

Take two more pieces of the same bare copper wire each about 20 cm. long. On one end of each piece of wire bend a hook and then attach the two pieces of wire to a ruler or similarly shaped piece of wood, as in the diagram below.

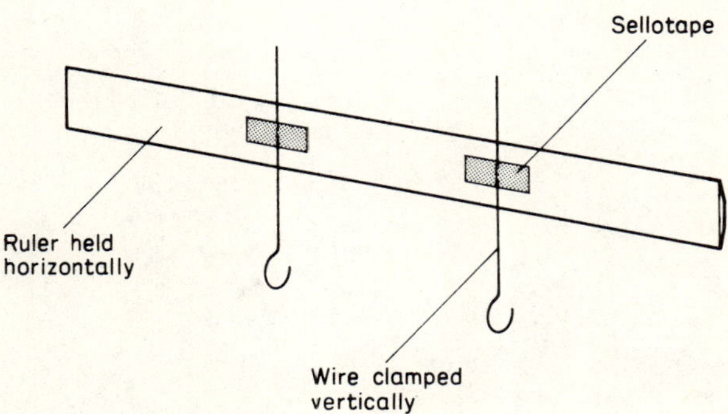

Attach the swing to the above wire as indicated.

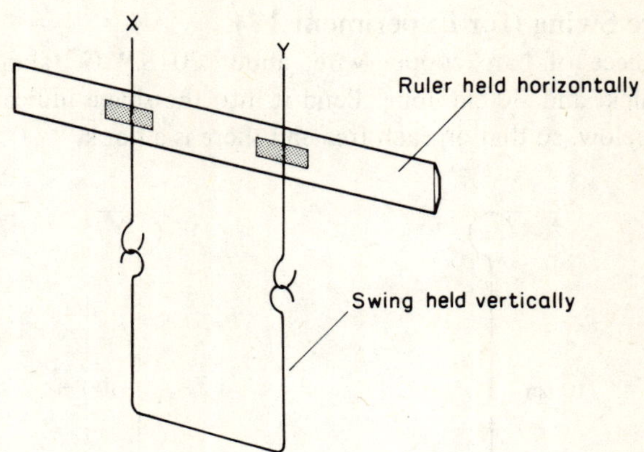

The electrical connections are to be made at the points X and Y.

Section Four

Additional background information

The Nature of Matter

If we take a large piece of copper A and divide it into two equal parts B and C, we find that the physical and chemical properties of A, B and C are all the same. The only difference being that B and C are smaller pieces than A.

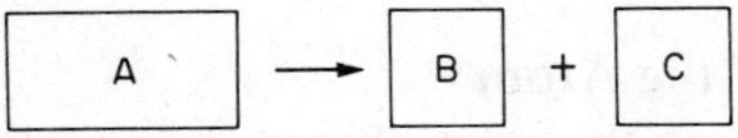

If we now take the piece B and divide it into two equal parts D and E, we again find that A, B, C, D and E have the same properties other than a reduction in the size of the pieces D and E.

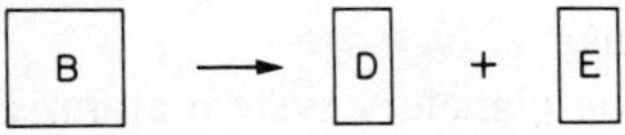

This process of dividing into two can be repeated a great many times, but eventually a piece — or particle — will be produced which, when we attempt to divide it, will not produce two halves having the same properties as the original piece of copper. This particle is called an ATOM.

If we repeated this procedure of division with carbon di-oxide, we would ultimately not produce a carbon di-oxide atom, but we would produce a carbon di-oxide *molecule*. This molecule is in fact made up of one carbon atom and two oxygen atoms.

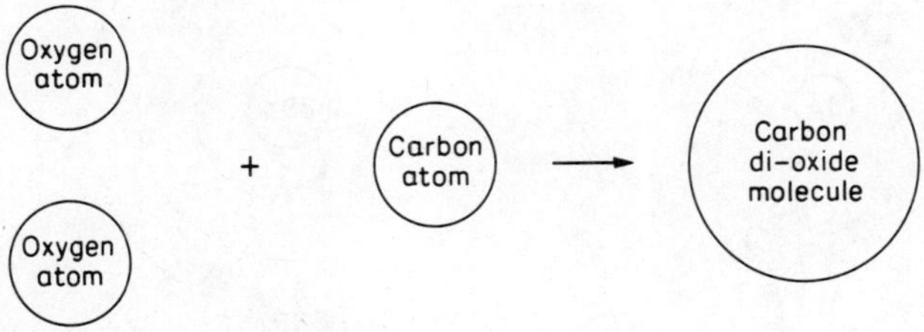

Molecules are thus made up from atoms. Another example is an ammonia molecule which is composed of one nitrogen atom and three hydrogen atoms.

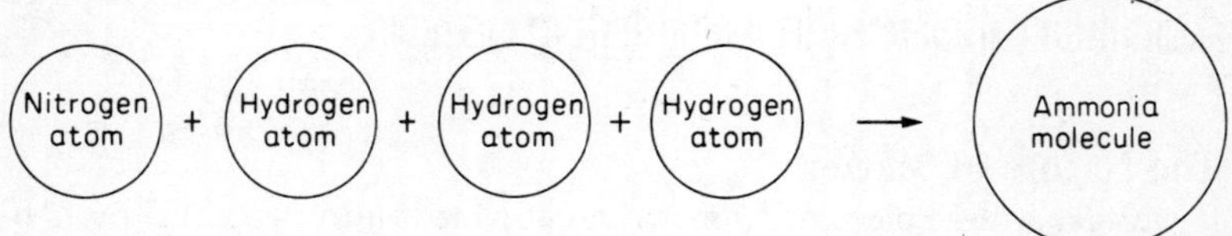

Substances are thus made up of either a collection of atoms, as for example copper and iron, or a collection of molecules, as for example carbon di-oxide and ammonia.

The Structure of the Atom

The early theories of the atom likened it to a small, solid, ball-like object, but more modern theories suggest that it can be pictured as a miniature planetary system. It should be stressed that the planetary system approach is really only useful to help visualise the structure of the atom and other modern theories explain the structure in terms of mathematical equations.

At the centre of the planetary system atom is a central core called the *nucleus*. The nucleus contains *protons* — these have a positive charge — and around it in orbits are tiny particles called *electrons* — these have a negative charge.

For the atom to have no charge, i.e. to be neutral, there must be the same number of protons in the nucleus as electrons in the orbit. Thus a hydrogen atom contains one proton in the nucleus, and one orbiting electron, whereas copper contains 29 protons in the nucleus and 29 orbiting electrons.

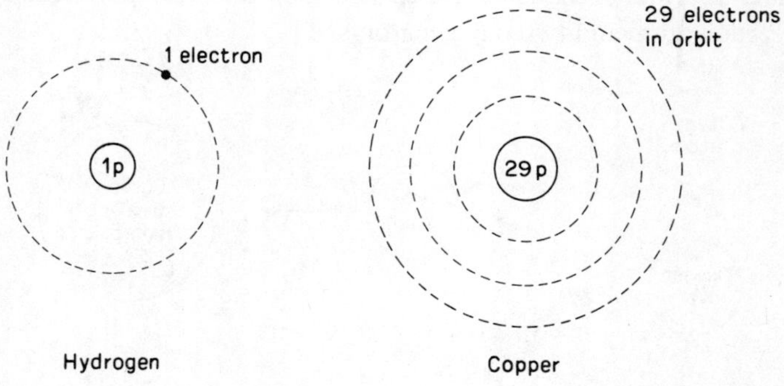

An atom is essentially empty space. This can be visualised quite easily if one imagines the nucleus of a hydrogen atom to be a football, and the electron to be a pin-head. The radius of the electron orbit would then be approximately 2 miles.

The protons and electrons have opposite charges and hence will attract one another. The force of attraction does, however, depend upon how far apart they are. The further apart they are, the weaker is this force of attraction. Consequently those electrons in the outermost orbit are only weakly attracted to the nucleus and can be removed from the atom comparatively easily.

If we remove an electron from an atom, i.e. we remove a negative charge, we leave the atom with more protons than electrons, and thus the atom will be positively charged. Alternatively, by adding an electron, we have an excess of negative charge and so the atom becomes negatively charged.

Since the electrons are in orbit around the nucleus they are far easier to remove than the protons which are in the nucleus, and so the charge of the atom is determined by the excess of electrons — in a negatively charged atom — or the deficiency of electrons — in a positively charged atom. Thus substances can be charged by rubbing them, since electrons will either be added or removed from the atoms of the substance.

The Nature of Electricity

Although a piece of copper appears to be a solid, there is between the atoms a great deal of empty space. Approximately one tenth of the space is occupied by an atom and nine tenths is empty space. The copper atom has one orbiting electron so far away from the nucleus that it can almost be considered to be free. It can in fact break away from the atom, move to the next atom and then move on after associating itself with that atom for a short time. Pictorially we can visualise the motion of this electron along a series of copper atoms thus:

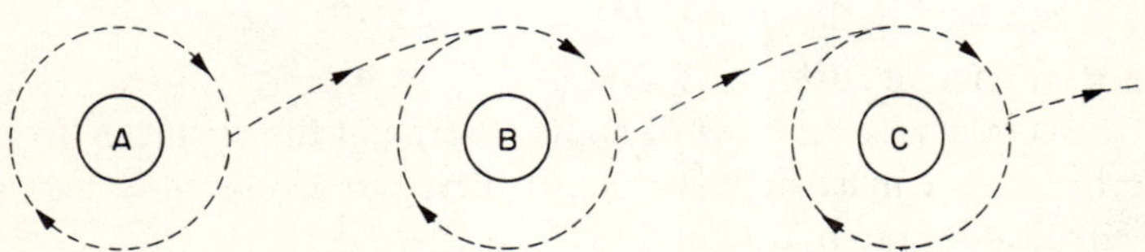

The outer electron orbits around A and then moves to B. After making a few orbits of B it moves to C and so on. Since copper is a three dimensional solid the electron could have moved in any of the three directions. This electron can then be regarded as a free electron. Since the outermost electron of each copper atom can be considered to be free, we have within copper a large number of free electrons — on average one per copper atom — moving in all directions between the copper atoms. The drift of these electrons is, on average, equal in all directions and so the copper remains uncharged.

Suppose that a battery is connected to the ends of a piece of copper wire. The positive plate of the battery attracts these free electrons, and so the electrons all move in one direction along the wire towards the positive plate. This then constitutes an electric current, i.e. it is the controlled flow of free electrons along the wire.

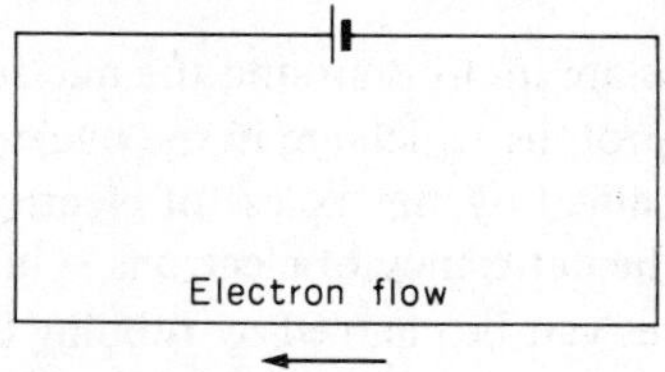

It should be noted that the electron flow is from negative to positive whereas we have said that conventional current flow is to be regarded as flowing from positive to negative. The reason for this is that the direction of current flow in a circuit was decided upon before electrons were discovered. For various economic reasons we retain current flow as being from positive to negative.

From this discussion it follows that substances having free electrons in them will allow electricity to pass through them and these substances are called conductors. Substances not having free electrons cannot pass a current and are called insulators. Thus the difference between conductors and insulators is due to the presence or absence of free electrons associated with the atoms of the substance.

The Electric Circuit

Since electricity is the flow of electrons around the circuit, it follows that the circuit must be complete. If there is a gap or an insulator in the circuit, the current cannot flow.

The electrons, in passing through the wire, do work and hence produce heat. Thus when an electric current flows in a wire, heat is produced. The greater the current flowing, the greater will be the heat produced. This is the reason why a fuse functions in a circuit. A fuse is simply a thin piece of wire which has a fairly low melting point. If too much current flows through the circuit a great deal of heat is produced which causes the wire of the fuse to glow red hot and then melt. This breaks the circuit and prevents the current flowing, thus protecting other possibly expensive electrical equipment connected in the same circuit from being burned out.

Magnets

If a bar magnet is freely suspended it will always settle pointing magnetic North and South — this differs slightly from geographical North and South.

The explanation of the earth's magnetic field is by no means clear, but its effect is as though inside the earth there is a tremendously strong bar magnet, as indicated in the diagram.

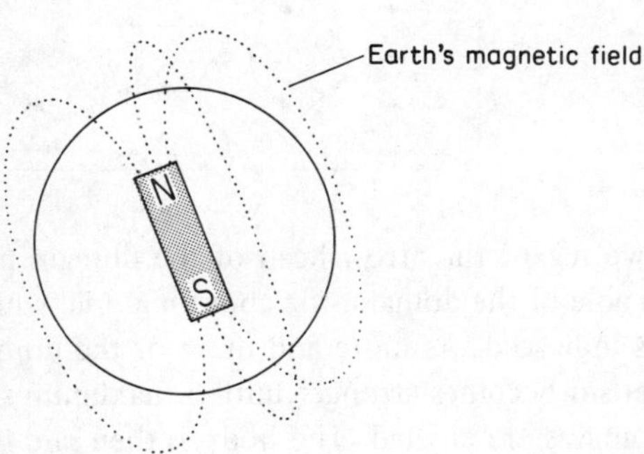

This magnetic field pulls the suspended magnet into the N-S position. A suspended light bar magnet is in fact a compass; care must however be taken over the method of suspending or pivoting the magnet.

Substances which exhibit these strong magnetic properties are called ferromagnetic since the property was first of all discovered in lodestone (an iron oxide ore). Iron and steel are the most common ferromagnetic substances, and also they show the largest magnetic effect, but in addition cobolt and nickel are also ferromagnetic.

Magnetic Domains

A simple theory of magnetism is to imagine that within the ferro-magnetic substances are these small magnetic domains which can be likened to small bar magnets. In an unmagnetised piece of ferro-magnetic material they form closed loops as indicated in the diagram.

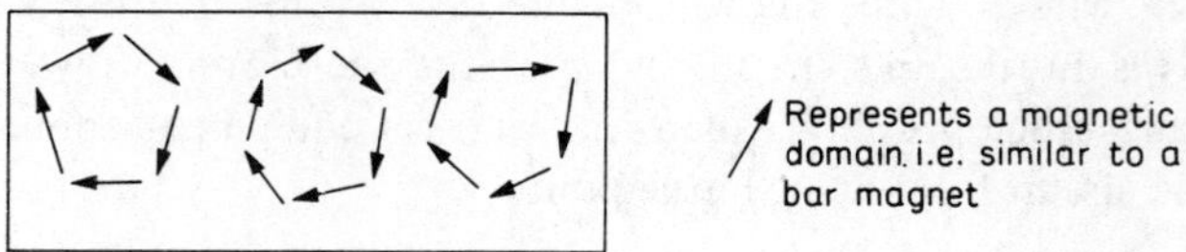

Since these closed loops are formed, the substance will, overall, be unmagnetised. Suppose that the material is subjected to a magnetic field, as for example when stroked by another magnet. This has the effect of aligning some of these domains thus:

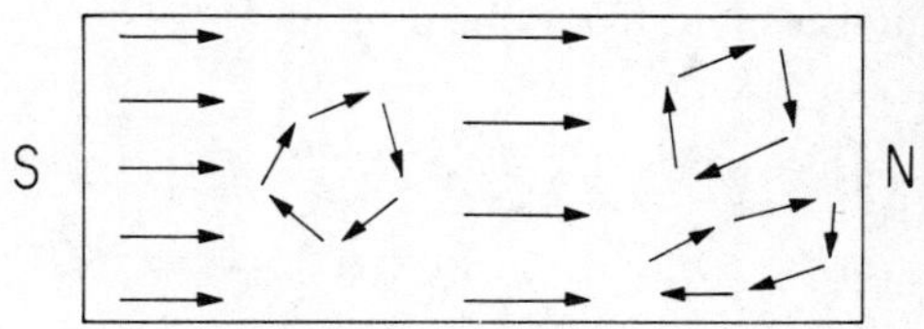

If we regard the arrow head of the domain picture to represent the North pole of the domain, the above material will be weakly magnetised N-S as indicated. As more and more of the domains are aligned, so the magnetism becomes stronger until a maximum state is reacted when all the domains are aligned. The body is then said to be magnetically saturated.

Electro-magnetism

Whenever an electric current passes through a piece of wire, it produces a magnetic field. If the wire is imagined straight and vertical the magnetic fields produced will be circular and horizontal. The diagram below shows the magnetic field about one point in the wire, and it should be realized that all points have similar fields produced about them.

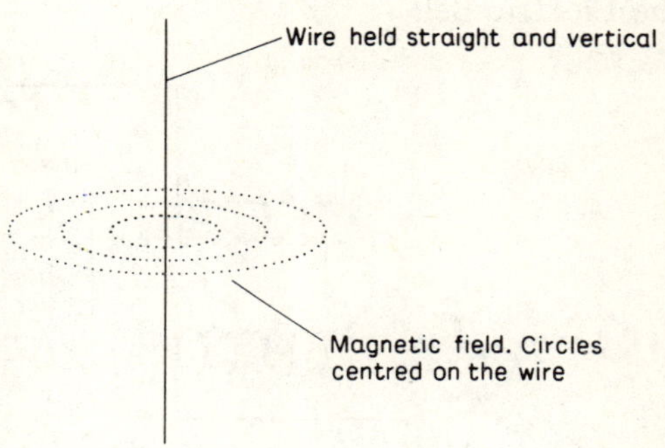

The direction of this magnetic field can be forecast by use of the corkscrew rule. Point the first finger of the right hand in the direction of the current flow — conventional current flows from positive to negative — and rotate the hand as though working a corkscrew. The direction of the thumb will then give the direction of the magnetic field. This is illustrated in the following plan diagrams i.e. the wire is imagined at right angles to the paper.

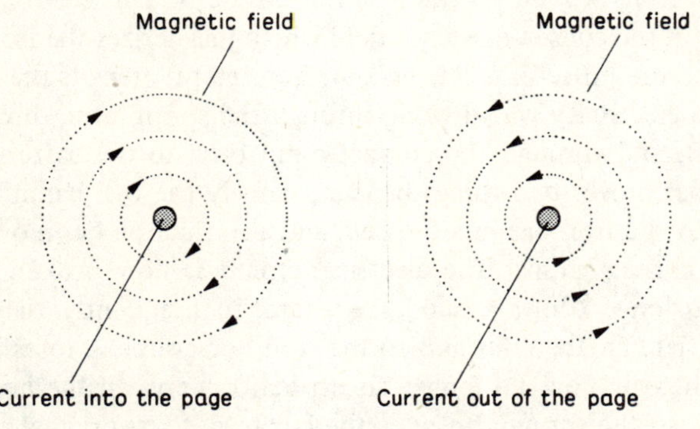

Knowing the direction of the magnetic field should enable an explanation of the deflection of the compass needle in Experiment 15 to be made.

The Electric Bell

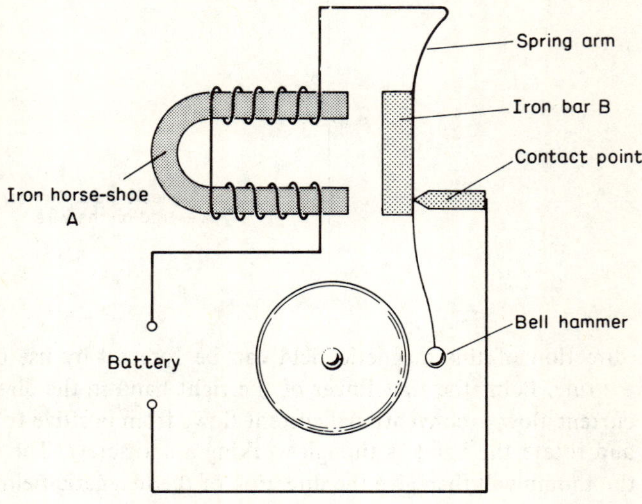

One interesting application of electro-magnetism is the electric bell, illustrated in the diagram. An electrical current flows from the battery around the coils wound on the iron horse-shoe A. This electric current produces, in the coils, a magnetic field which magnetizes the iron horse-shoe. Since the horse-shoe has become a magnet it attracts the iron bar B, but the electricity was flowing through the spring arm, through the iron bar B, and through the contact point back to the battery. When the iron bar moves two things happen, namely the bell hammer being attached to the iron bar hits the bell, and also the iron bar moves away from the contact point. The electrical circuit is now broken and the current no longer flows around the circuit, consequently there is no magnetic field in the coils and so the iron horse-shoe A loses its magnetism. There is now no longer an attraction between the horse-shoe and bar B, so the spring arm pulls the bar back to its original position. In doing so, contact is made with the contact point, thus completing the electrical circuit and beginning the whole procedure all over again. The electric bell is therefore the continual switching on and off of an electric current, the switch being worked magnetically, and the bell being hit once each time the current is switched on.

The Electric Motor Effect (Electro-dynamics)

We have seen that when two magnets, i.e. two magnetic fields, are brought close together a force is set up between them and either attraction or repulsion occurs. Since a magnetic field is produced when an electric current is passed through a wire, it seems reasonable to imagine that if we place this wire in a magnetic field, then on passing a current through the wire a second magnetic field will be produced, which should interact with the first magnetic field and produce a force. This is in fact what happens in Experiment 17.

The direction of this force and the resultant movement can be forecast using Flemmings Left-Hand Rule. If you look at the results of Experiment 17, on page 38, you should see that the magnetic field between the two magnets (remember the field goes from North to South) and the direction of the current along the cooking foil (from positive to negative) are at right angles as in the diagram below.

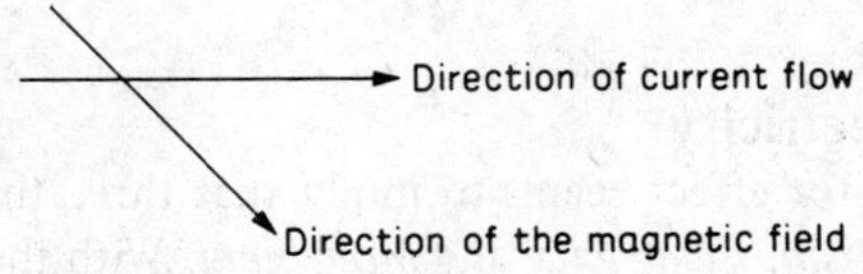

The resulting force, as illustrated by the movement of the cooking foil, should be seen to be at right angles to both the magnetic field and the current flow as illustrated below:

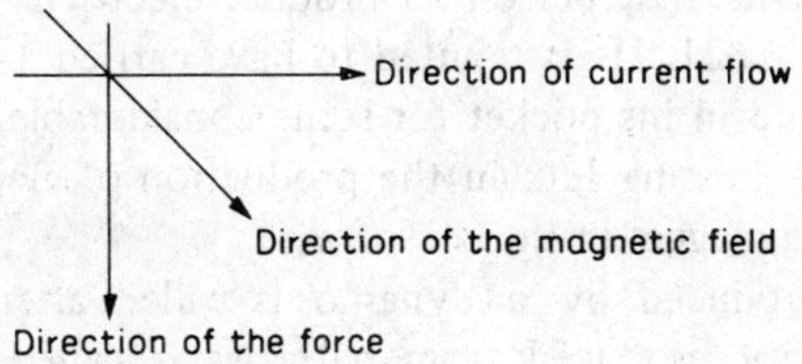

This can be remembered or forecast by holding the thumb, first and second fingers of the left-hand all at right angles to one another. The first finger should be pointed in the direction of the magnetic field, the second finger in the direction of the current flow, and the thumb will give the direction of the force. This is known as Flemmings Left-Hand Rule and is illustrated in the diagram below.

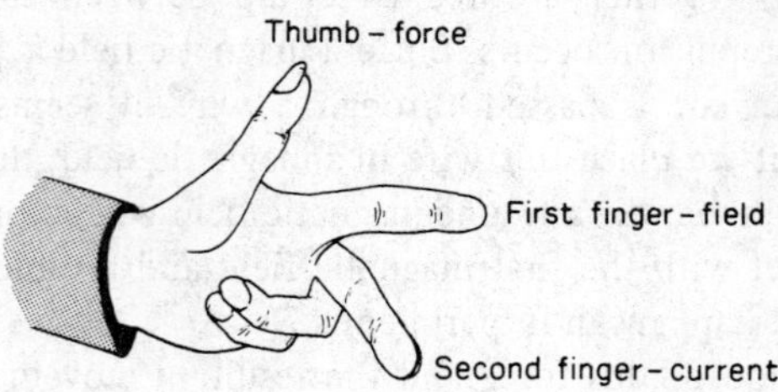

The fact that electricity can produce movement in a magnetic field is very important since electric motors play a very large part in our lives. They are used in such appliances as washing machines, refrigerators, vacuum cleaners, hair dryers, and starter motors for cars. In fact there are very few electric appliances that do not have an electric motor in them.

Producing Electricity

The electric motor effect seems to imply that three things are present, namely, *Magnetism, Electricity* and *Movement*. With the motor, magnetism and electricity were present and movement was produced. It would thus appear a possibility that if magnetism and movement were present then electricity could be produced. This is in fact the case as is illustrated in Experiment 20, and thus leads to the practical application, namely the dynamo.

Faraday was the first person to produce electricity in this way on New Year's Day 1821. He is reputed to have carried a bar magnet and coil of wire round in his pocket for some considerable time before he realised that the missing link in the production of electricity was the movement of the magnet in the coil of wire.

Electricity produced by a dynamo is called alternating current, whereas electricity produced from batteries is called direct current. The difference between these two types of electric current is discussed in most electricity text books, but is considered somewhat complicated for the young pupil.

Chapter Two
Heat

Section One

Experiment 22
Expansion of solids, liquids and gases

Objectives
a To observe that solids, liquids and gases expand on heating.
b To consider some applications of these effects.

Apparatus ·per group
Candle or calor gas heater, vegetable dye, supply of quite hot water, hair-dryer. Screw and eye, and liquid-in-vessel apparatus (see apparatus construction, page 96, for these two items).

Procedure
i Issue the screw and eye apparatus to each group. If you consider it safe to allow the children to have a source of heat, e.g. a candle or calor gas heater, then one should be issued to each group, otherwise the source of heat will have to be kept on the front teaching bench.

Allow the class to establish that the screw can pass through the eye quite easily.

Suggest to the class that they should heat up the screw and see if it still passes through the eye. Since the class should find that the screw will not now pass through the eye, they should be asked why it does not and how it can be made to.

The screw has obviously expanded on being heated and to allow it to pass through the eye, the eye can be heated and expanded, or alternatively the screw can be cooled down again by placing it in cold water.

This experiment thus shows that a metal on being heated will expand.

ii To show that liquids expand on heating, the liquid-in-vessel apparatus should be used, and this apparatus should be issued to each group.

The class should be encouraged to mark the level of the liquid in the tubing using either rubber bands or sellotape.

The apparatus should then be immersed quickly into hot water and the class should be asked to observe and record what happens to the level of the liquid.

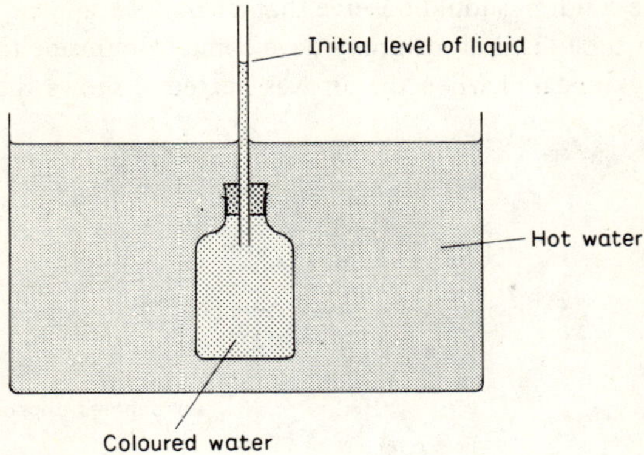

They should observe that the level of the liquid in the tube rises, showing that, on heating, the liquid has expanded.

iii To show that gases expand on heating, use the same apparatus as in the previous experiment after the coloured water has been removed.

Tell the class that they should hold the apparatus upside down so that the tubing is *just* below the surface of the water. They can then warm the flask, i.e. the air inside it, with their hands and observe what happens.

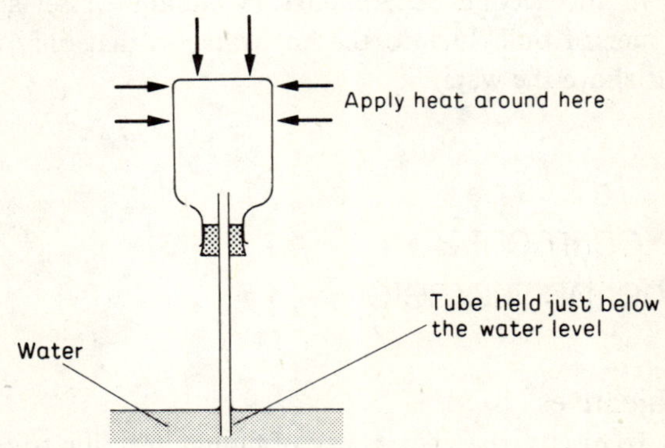

Heating the air by this method may not be satisfactory in which case the flask should be heated with a hair-dryer.

The children should observe that air bubbles appear from the bottom of the tube. This air can only have come from inside the flask and since it only appeared when the air was heated it shows that air expands on heating.

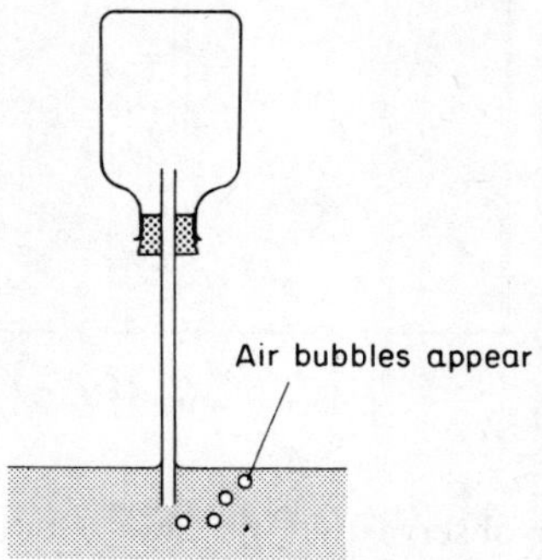

iv There are many applications of expansion in everyday life (see additional information on page 100) and they should be discussed at this stage.

Notes
1 In the first experiment the screw should be heated to quite a high temperature and thus care must be taken to see that the children do not burn themselves.
2 In the second experiment the liquid-in-vessel apparatus should be immersed quickly into the hot water so that only the tube is sticking out above the water.

Experiment 23
The thermometer

Objectives
a To discuss the uncertainty of human reaction to temperature.
b To use a simple mercury-in-glass thermometer to measure temperature.
c To appreciate the fixed points of a thermometer.

Apparatus per group

One mercury-in-glass thermometer, three large bowls, pure ice, pure water. Method of boiling water, e.g. an electric kettle.

Procedure

i We tend to say that it is warm, cool or cold and so on, and yet these statements are only relative and depend upon our reaction to the temperature we are in contact with. It is essentially the way our body is reacting to this temperature and it is possible for parts of our body to react differently to the same temperature. This can be illustrated by this first experiment.

The children should be issued with three bowls of water, the first containing hot water, the second cold water and the third cool water. They should be told to put one hand into the hot water, and the other hand into the cold water. After about twenty seconds they should put *both* hands, at the same time, into the cool water and be asked to note their reactions.

They should note that the reaction of the two hands is different although they are being exposed to the same temperature, thus emphasising that our bodies are not particularly good or reliable indicators of temperature.

ii Temperature is merely how hot a body is, and a thermometer gives us a number to this effect. Thus a temperature of $20^{O}C$ is hotter than a temperature of $10^{O}C$, but it is not *twice* as hot. The most commonly used scientific thermometer is the mercury-in-glass thermometer.

Each group should be given a thermometer and they should note that it is only a length of mercury inside a glass tube. As the temperature rises the mercury expands and moves further up the hole inside the glass tube. At a temperature of, for example, $25^{O}C$ the mercury level will always be at the same point in the tube and hence that point on the thermometer can be labelled $25^{O}C$.

The children should be given practice in reading temperatures and should be asked to measure the room temperature, the temperature of water in a fish tank, the temperatures of the hot and cold water in the wash basin and the temperature outside the classroom.

iii There are two very important temperatures, namely $0^{O}C$ and $100^{O}C$.

The children should be asked to find the temperature at which *pure* ice melts. They should find that this is in fact $0^{o}C$. They should also be asked to find the temperature at which *pure* water boils — care must obviously be taken to see that they do not scald themselves. This temperature they should find to be $100^{o}C$.

iv The children should then be asked how they could begin to put the markings onto a thermometer if they were given one with none on. Obviously if pure ice and pure water were available then the $0^{o}C$ and $100^{o}C$ marks could be made and the space between them divided into 100 equal divisions. This process is called *calibrating* a thermometer. The two temperatures $0^{o}C$ and $100^{o}C$ are called the *fixed points* and are thus very useful in calibrating thermometers.

Notes
1 When finding the melting point of pure ice make sure that the ice is melting, i.e. that it is really ice in contact with water formed from the melting ice.
2 When finding the boiling point of pure water, make sure that as much of the stem as possible of the thermometer is under water, i.e. all the thermometer should really be at the temperature of the boiling water. This is obviously a dangerous experiment.
3 Certain things do alter the temperature of the fixed points, as for example adding an impurity such as salt. This can be tried as an experiment by the children.

Experiment 24
To make a simple thermometer

Objectives
a To make a simple thermometer.
b To calibrate the thermometer.
c To use it to measure certain temperatures.

Apparatus per group

The liquid-in-vessel apparatus (see apparatus construction, page 96), industrial alcohol or water, vegetable dye, a mercury-in-glass thermometer.

Procedure

i In this simple thermometer the liquid used will not be mercury, but can be industrial alcohol, or if that is not available then water will answer the purpose. The alcohol or water will need to be coloured to make it easy to see and so any vegetable colouring material — red ink will in fact do — can be used.

The thermometer consists of a bulb or reservoir of liquid into which is dipped a long thin narrow tube, i.e. the liquid-in-vessel apparatus. The liquid must completely fill the space in the reservoir — no air bubbles should be present — and partly fill the tube. As the temperature changes the volume of the liquid changes, thus the liquid level in the tube will change. Consequently, to convert this apparatus into a thermometer, a scale needs to be fastened behind the tubing, as indicated in the diagram.

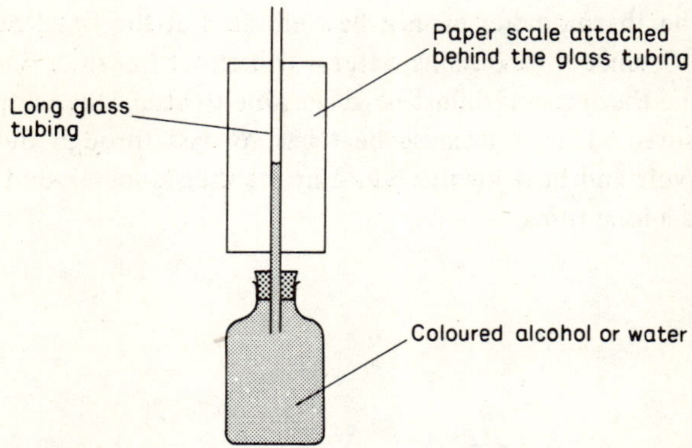

The children should first of all construct their own thermometers and then be asked how they think they can calibrate it.

ii In Experiment 23 the two fixed points 0°C and 100°C were discussed, but obviously these two are not suitable when calibrating this home-made thermometer. The children should be encouraged to discuss why these two fixed points are unsuitable.

A reasonable method is to calibrate it against a mercury-in-glass thermometer. To do this the home-made thermometer and the mercury thermometer should be placed in the same container of cold water. The temperature indicated on the mercury thermometer should then be recorded on the scale of the home-made thermometer. This procedure should then be repeated for different temperatures of the water, ranging from say 10°C to 80°C.

Throughout these processes, care must be taken to see that the whole of the reservoir of the home-made thermometer is at the same temperature, which means that it must all be covered by the water used in the calibration. A bucket could be used to hold the two thermometers and the water during this part of the experiment.

iii Having calibrated the home-made thermometer, it can now be used to measure temperatures. The children should be encouraged to use it to measure temperatures such as the room temperature, the temperature outside the room, the temperature of the hot and cold water in the wash basins and so on.

Notes
1 The thermometer cannot be calibrated at the fixed points 0°C and 100°C, since if it contains water it will either freeze or boil away.
2 The thermometer must be given time to attain the temperature being measured. This is because heat has to pass through the walls of the reservoir and heat up the liquid in the thermometer, and this may take quite a long time.

Experiment 25
To show convection currents in gases

Objectives
a To show that heat can be transmitted through gases by convection.
b To consider some applications of this.

Apparatus per group

Candle, source of smoke, light plastic bag. Mine-shaft apparatus, paper convection current detector and hay-box (see apparatus construction, page 97, for these three items).

Procedure

i   To show the existence of convection currents in gases use the mine-shaft experiment. This is probably best demonstrated by the teacher in front of the class.

Light the candle and ask a child to hold his hand at the point X and then at Y, and say what he observes.

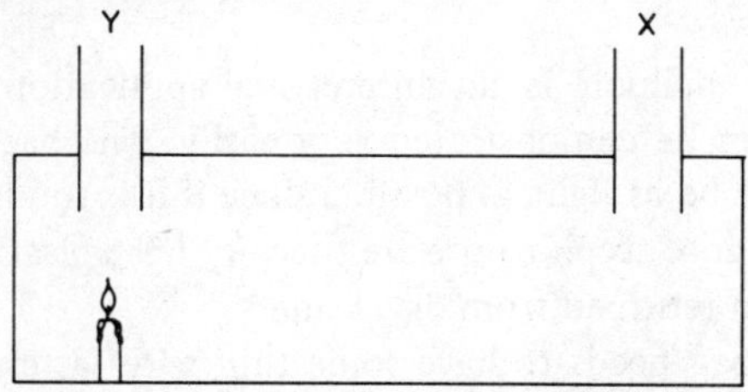

He should feel that X is quite cold but that Y is hot, i.e. heat from the candle is arriving at the point Y but not the point X.

This can be investigated further by bringing a source of smoke, e.g. a smoking oily cloth, to the point X and watching the passage of the smoke. It should follow the path as indicated in the diagram.

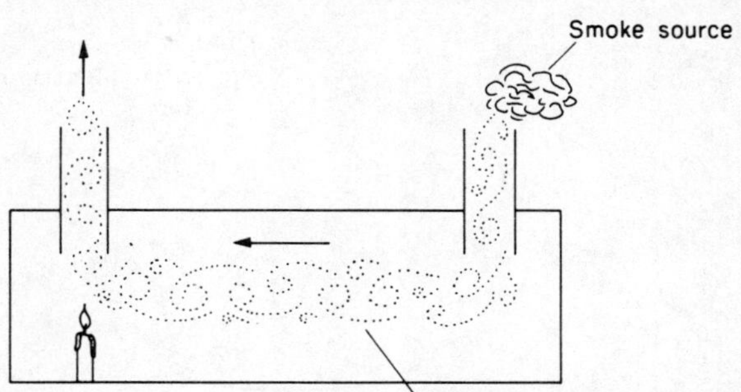

This experiment thus shows that the heat from the candle has caused the air to move. The point Y was hot showing that the hot air had risen, allowing cold air from X to flow in. This flow of air is shown by the passage of the smoke.

Consequently, the heat produced by the candle in the air is carried by the movement of the air by what are called *convection currents*.

ii Convection currents in air can be detected using the paper detectors described in the apparatus construction section, page 97.

Each group should be given a candle and a paper detector and asked to trace the path of the convection currents produced by the lighted candle. Some groups could investigate what happens if two lighted candles are placed side by side.

In rooms containing radiators it may be possible using these paper detectors to follow the path of the convection currents in heating the room.

iii The hot air balloon is an interesting application of convection currents and can be demonstrated very easily. The bag containing the hot air needs to be as light as possible since if it is too heavy it will not rise at all. One good type to use are the very light plastic bags in which clothes are often returned from drycleaners.

The plastic bag needs to have some thin wire fastened to the open end to provide a rigid shape to it. The open end is then held over a source of heat such as a candle — be very careful not to melt the plastic — and when the air inside the bag is warm enough the bag should be released. The ascent into the air is quite spectacular.

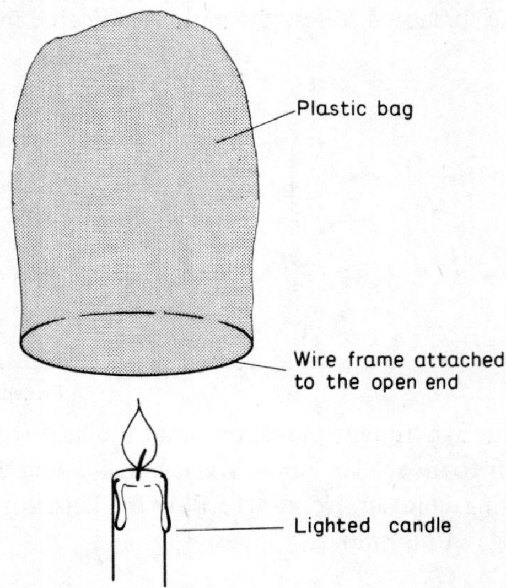

iv Another interesting application, and one the children should set up, is the *hay-box* method of keeping things warm. If, for example, you are camping and wish to have reasonably warm porridge for breakfast, but do not want to cook it in the morning, it can be cooked at night and put into this hay-box. As the name implies it is merely a box with fairly loosely packed hay inside it, inside which is placed the hot container.

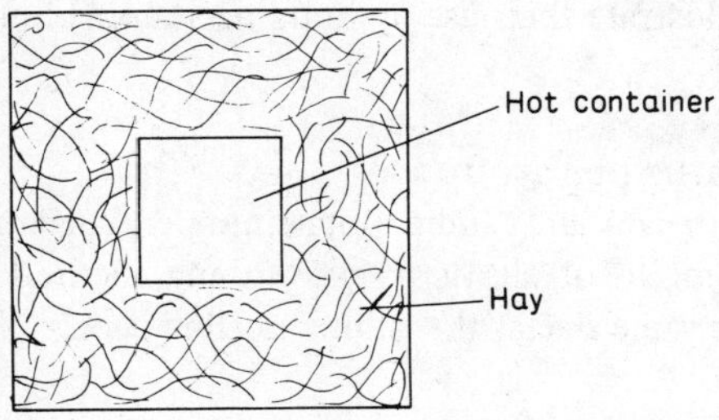

The reason that little heat is lost to the outside, thus keeping the container hot, is that there are tiny pockets of air formed in the hay and the convection currents are confined to their own tiny pockets. Since these are small, the hot air merely circulates inside each pocket and does not escape to the outside.

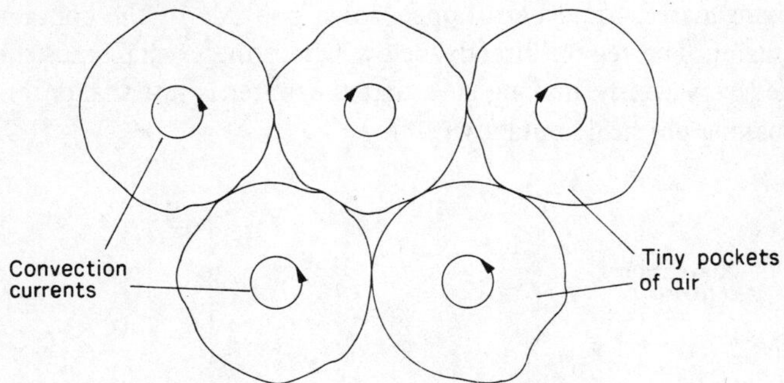

The children should be encouraged to set up a hay-box, leaving something hot in it overnight.

Notes
1 Make sure that the hay in the hay-box is not packed too tightly.
2 If hay is not available then straw, wood shavings or a loosely knitted, woollen object will act as useful substitutes.

Experiment 26
Convection currents in liquids

Objectives
a To show the existence of convection currents in water.
b To illustrate their use by building a model hot water system.

Apparatus per group
Beaker or jam jar, candle, potassium permanganate, two jars with corks
to fit, lengths of plastic or glass tubing, method of drilling a hole in the
cork the same size as the tubing, sealing wax.

Procedure
i The children should be issued with a beaker or jam jar full of
water, a source of heat and a few crystals of potassium permanganate.

The container of water should be left for a few minutes until there
is no movement of the water. Very gently a single crystal of potassium
permanganate should be dropped down one side of the container into
the water. The region directly below where the crystal lands should be
heated very gently making sure that the water is not shaken or moved
by making physical contact with it.

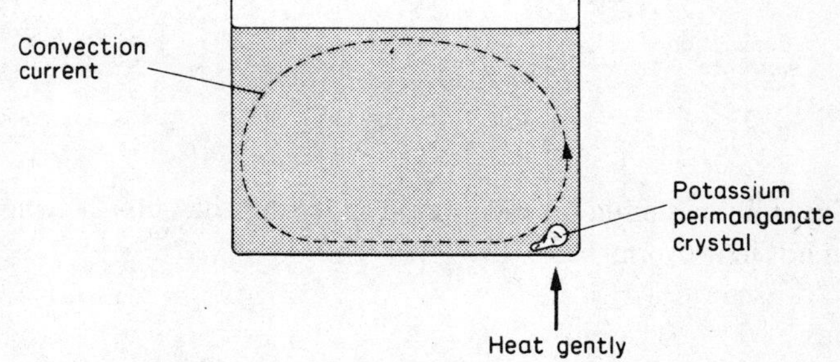

The potassium permanganate will dissolve giving a purple dye which will trace out the convection currents. The water must be heated gently otherwise the colour spreads through the water far too quickly for anything to be observed.

Convection currents in gases or liquids can thus be summarised very easily by saying that the heat is transported through a liquid or gas by the liquid or gas actually moving and taking the heat with it.

ii One important application of convection currents in liquids is the domestic hot water system. The details that follow are an account of the workings of the system and instructions for building a model.

iii The water is usually heated in some form of boiler by a flame produced by oil, gas or coal. Thus the hot water in the boiler rises and if a pipe outlet at the top of the boiler exists, then hot water will flow up this. Consequently cooler water is allowed to flow into the boiler at the bottom.

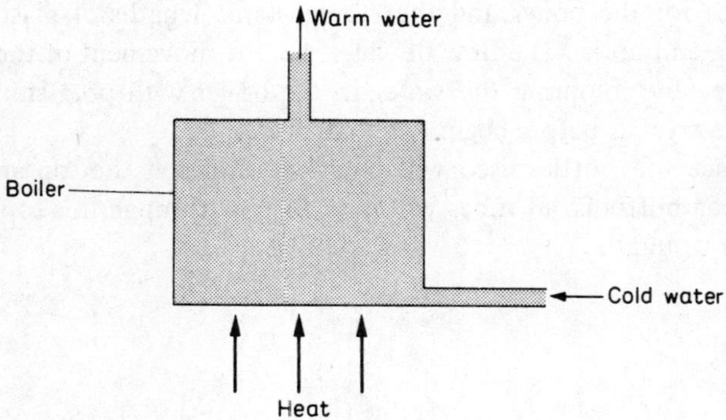

iv This pipe from the top of the boiler passes up to a storage tank (usually in the airing cupboard), and should enter this tank at the top. The water in this storage tank will inevitably cool down, and will thus sink to the bottom of the tank. Consequently water from the bottom of the storage tank should be returned to the bottom of the cylinder.

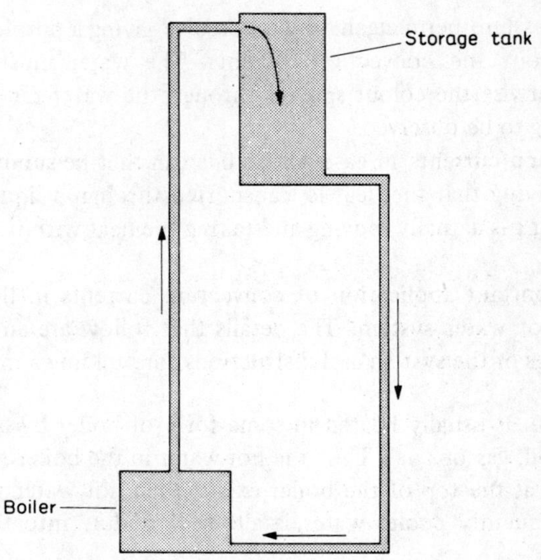

v A model of the situation as developed so far can now be built using bottles for the boiler and the storage tank, lengths of plastic or glass tubing and corks. The flow of water, i.e. the movement of the heat, can be seen by colouring the water in the boiler with potassium permanganate crystals before beginning to heat it.

Since the bottles used will only have holes at the top and not the sides or bottoms, all tubes will have to pass through this top. Thus the boiler would be:

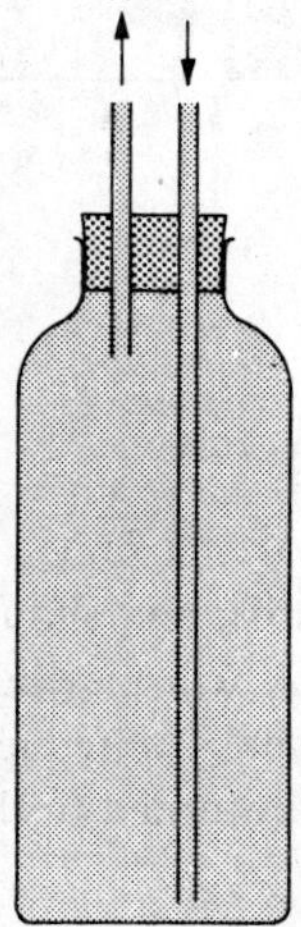

Water would rise through tube X, corresponding to the top of the boiler, and return through tube Y. Since tube Y ends at the bottom of the jar it is the same as bringing the water directly into the bottom of the boiler.

The set-up for the storage tank could be very similar.

vi The model, so far, is very simple since it does not allow water to be drawn from the system. If water is to be removed, then fresh water from outside must be allowed in. The water will be cold, and thus must be let in at the coldest part of the circuit which is just where the water is passing back into the boiler.

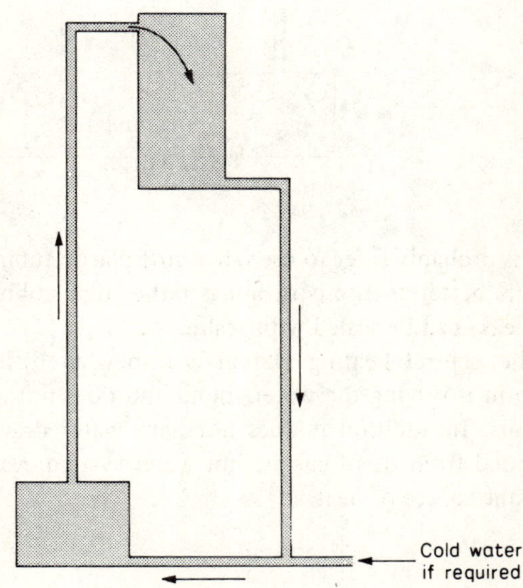

Water that will be drawn off needs to be as hot as possible and should either be taken from the top of the storage cylinder or the pipe carrying the water from the boiler to it.

vii Consequently by adding in the appropriate tubing a fuller scale model of the domestic hot water system can be built.

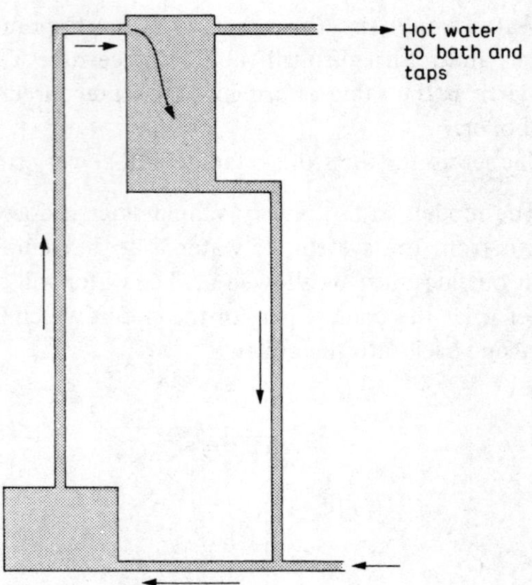

Notes

1 It is probably safer to use fairly stiff plastic tubing rather than glass.

2 It is better to use cork bungs rather than rubber bungs in the jars. Any leaks can be sealed with sealing wax.

3 The central heating system is somewhat different since it has a pump in it driving the water round and does not rely upon convection currents. In addition it does not have water drawn from it and so is separated from the domestic hot water system even though it may use the same source of heat.

Experiment 27
Heat transfer by conduction and radiation

Objectives

a To show that heat can be transferred by conduction.
b To show that heat can be transferred by radiation.
c To consider some applications of these.

Apparatus per group

Two similar shaped bars, one of metal the other glass. A long, narrow glass container. Two thermometers, black and silver paint.

Procedure

i Experiment 26 showed that heat can be transferred by convection, i.e. the body actually moving and carrying the heat with it, but obviously this cannot be the case with solids. Heat, however, is transmitted through solids as is demonstrated by the next experiment.

ii The class should be supplied with a metal rod, and a similar sized glass rod. They should hold one in each hand and place the free end of each rod into the same candle or calor gas flame. After a fair length of time they should be asked for their observations.

They should observe that the hand holding the metal rod feels warmer than that holding the glass rod. This means that heat has been transmitted down the metal bar and the glass bar, but more has passed down the metal bar.

This transmission of heat is called *conduction*. Metals are said to be good conductors of heat, whereas non-metals, being poor conductors, are called insulators. It should be noticed that good conductors of heat are good conductors of electricity.

iii Heat can also be conducted through liquids and gases, but in general they are very poor conductors and rely upon convection for transmission through them. The following experiment demonstrates how poor a conductor of heat is water.

The class should be issued with a long, thin glass container which they should almost fill with cold water. They should heat the water at the top of the container with a candle and hold the container at the bottom as in the diagram.

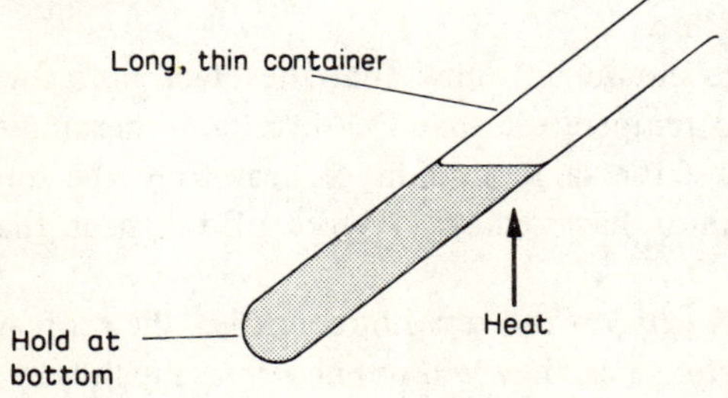

Throughout this experiment they should be told to watch the water at the top of the container, i.e. that in the region they are heating.

Eventually they should find that the water at the top will boil and yet they can still hold the container at the bottom quite comfortably.

This means that very little heat has been conducted through the water, showing that water is a very poor conductor of heat.

Convection currents have been eliminated by heating at the top. Since hot water rises for convection, the water at the top cannot rise and so no convection currents exist.

iv The main source of heat on the earth is from the sun, and yet this heat has to travel through empty space, which is called a vacuum. Since there is this empty space the heat cannot be transmitted by convection or conduction, and so there must be a third method of transmission. This method is called *radiation*.

Radiant heat has the ability to travel through a vacuum, through gases, through liquids and some solids, for example, glass. It also travels at a very high speed, in fact it has the same speed as light. This can be seen clearly if one observes a cloud passing in front of the sun on a fine day, when it should be noticed that the warmth from the sun stops at the same time as the light coming from the sun.

Radiant heat can be shown to have several interesting properties, one of which is illustrated by the following experiment.

v This experiment is probably best performed by the teacher as a demonstration since it can be left for an appreciable length of time while other work is being considered.

Set up two thermometers so that the rays from the sun fall equally on the bulbs of each thermometer. Paint one bulb black, and the other bulb silver and leave the two thermometers exposed to the sun for the same length of time.

Eventually it should be found that the black bulb thermometer is reading a higher temperature than the silver bulb thermometer, and yet they each receive the same amount of heat from the sun. Hence the black surface must have taken in more of the heat than the silver surface.

This explains why, in the very hot countries, the roofs of the houses are painted white, since they will not absorb as much heat as if painted a darker colour, and why we should wear white colours in summer to keep cool.

Notes

1 In the first experiment it is necessary to have the two bars approximately the same size.

2 The second experiment should end when the water at the top is boiling. If it is continued after this, heat will begin to be conducted by the glass and the child may burn his hand.

Section Two

Apparatus required and suppliers

The items listed below are sufficient to make one complete set of apparatus for the experiments indicated.

Experiments. 22, 23, 24, 25, 26 and 27

All the items listed may be obtained from a chemist or a hardware shop, except those otherwise indicated.

Source of heat — candles, calor gas heater, hair-dryer.

Screw eyelet — 3/8 inch diameter.

No. 6 wood screw.

Two pieces of soft wood, approximately 20 x 1 x 1 cm.

Ink bottle or thin-walled glass container.

A cork to fit the above bottle.

Glass or plastic tubing, between ½ and ¼ cm. diameter.

Vegetable dye.

Two mercury-in-glass thermometers, reading from $10^{\circ}C$ to $110^{\circ}C$.

Distilled water (obtainable from a garage).

Ice cubes.

Cooking salt.

Industrial alcohol or methylated spirits.

Shoe box, approximately 25 x 10 x 10 cm.

Plastic or glass sheet to fit the front of the above box.

Two glass or non-combustible tubes, approximately 2 cm. in diameter and 15 cm. long.

An oily cloth which should produce smoke when set alight.

Large light plastic bag (obtainable from a drycleaners').

Potassium permanganate crystals.

Identical metal and glass rods, approximately 1 cm. in diameter and 20 cm. long.

Sharp knitting needle.

Large cardboard box, approximately 50 x 50 x 50 cm.
Hay, straw or wood shavings to fill the above box.
Two jam jars with corks to fit.
Sealing wax.
A long narrow glass container, approximately 2 cm. in diameter and 20 cm. long.
Small supply of black and silver paint.

Section Three

Apparatus construction

**Screw and eye apparatus
(for Experiment 22).**
A suitable screw eyelet can be obtained from a hardware shop selling
curtain rail fittings, since they are often used in the ends of the plastic
covered cable holding up net curtains. The small ones are about 3/8
inches in diameter, and a suitable screw is a no. 6. The screw head
should just pass through the eyelet. Fasten the eyelet into one end of a
piece of wood, and the screw into another piece, as in the diagram.

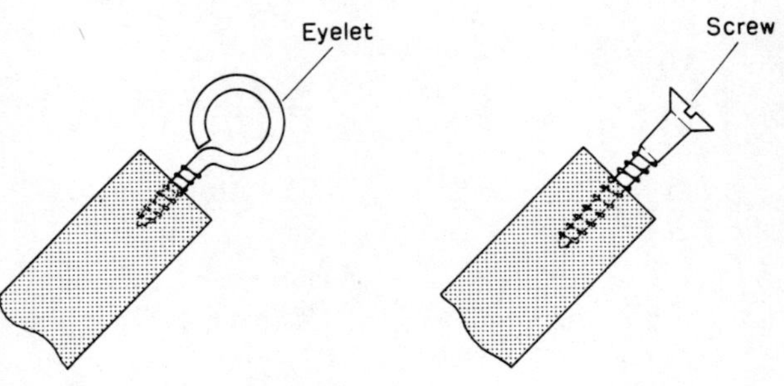

Liquid-in-vessel apparatus (for Experiment 22, 23 and 24).
To construct this apparatus a transparent container such as an ink
bottle with a tight fitting cork is required. Drill a hole through the
cork so that a piece of glass or clear perspex tubing will just pass through
the hole. The tubing should stand vertically and it may therefore be
necessary to arrange a wooden or cardboard support if flexible perspex
tubing is used. The tubing should be between ½ and ¼ cm. in diameter
and about 50 cm. long. The bottle and part of the tube should be filled
with coloured water — a suitable colouring agent is a potassium permanganate crystal or vegetable dye. Care must be taken to see that there
are no air bubbles in the top of the container just below the cork.

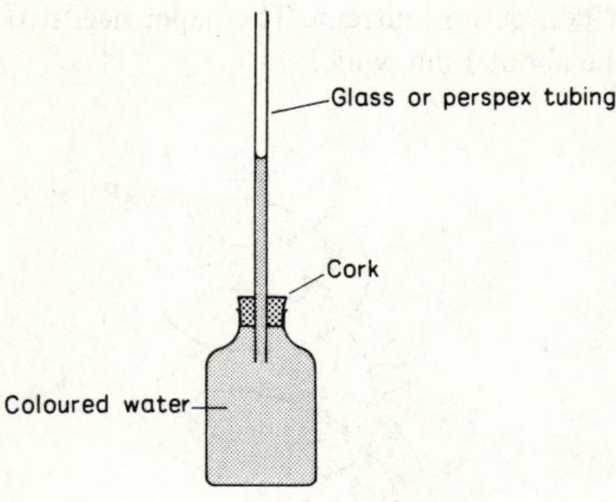

Mine Shaft Apparatus (for Experiment 25).

Remove the front from a shoe box — approximately 25 x 10 x 10 cm. — and replace it with a piece of glass or heat resistant plastic. Into the top of the box fix two glass or plastic tubes, approximately 2 cm. in diameter and 15 cm. long. Place the candle below one of the tubes and make sure that all joints are air-tight so that the only place that air can get into the box is through the tubes.

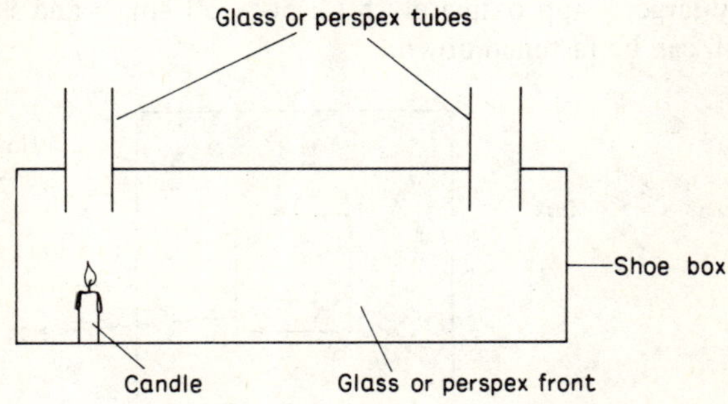

Detectors of Convection Currents in Air
(for Experiment 25).

Construct a paper spiral as in the diagram and balance it on the sharp point of a knitting needle. The spiral will be seen to spin when in the

stream of a convection current. The paper needs to be light but rigid
and should be about 1 cm. wide.

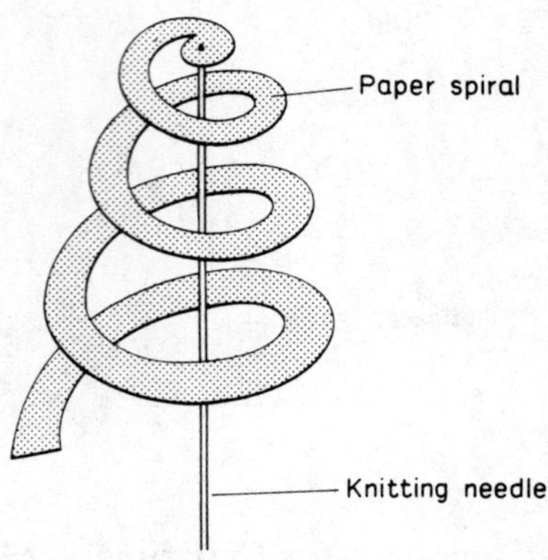

Hay-box (for Experiment 25).

A hay-box is merely a large box containing loosely packed hay, straw
or wood shavings, inside which is placed a hot body. The main aim is to
keep the body hot by preventing the heat escaping. The box should be
fairly large — approximately 50 x 50 x 50 cm. — and should have a lid
which can be fastened down.

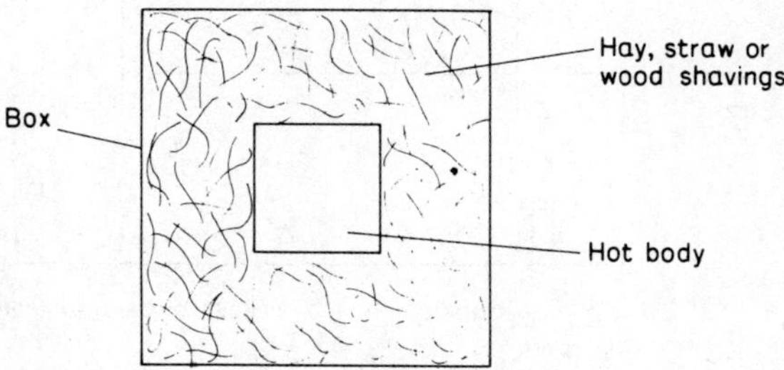

Section Four

Additional background information

Structure of Solids

In the previous section it was indicated that substances were made up of atoms or molecules, but no reference was made as to how they were arranged in the actual substances. The simplest explanation is to imagine that the atoms are arranged in regular arrays in the solid, the simplest being the cubic structure where the atoms are on the corners of cubes, as in the diagram below.

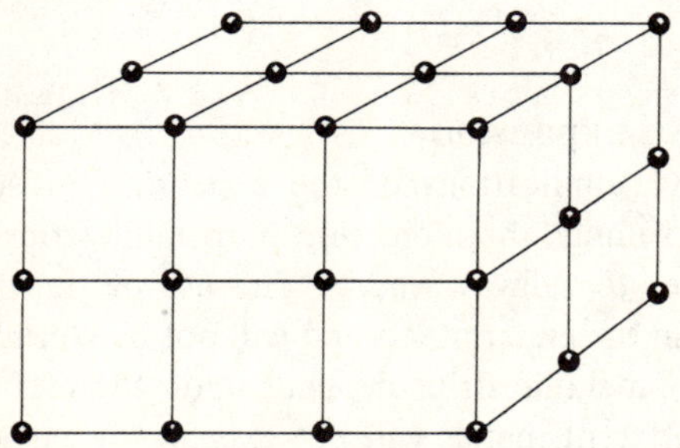

This regular pattern then repeats itself throughout the substance.

The atoms are not stationary but vibrate — in three dimensions — about these fixed points. When heat is applied, the atoms vibrate faster and further, in fact they take up more room and so the solid expands. When the solid cools down, the vibrations of the atoms decrease and so the solid contracts.

Structure of Liquids

The atoms and molecules of a solid are held in position by the mutual forces of attraction and repulsion between each other. If, however, they are given sufficient energy they will vibrate so much as to escape from their fixed points and the solid becomes a liquid.

As the temperature of the liquid increases so the speed of these relatively free atoms and molecules increases, causing them to move further. The surface boundary of the liquid is thus extended and so the liquid expands. The reverse effect happens if the liquid is cooled down.

Structure of Gases

If the liquid is heated still further an atom or molecule in the liquid will be given sufficient energy to overcome the attractive forces between itself and the remaining liquid molecules and will escape from the surface of the liquid. The liquid thus gradually becomes a gas.

If a gas is contained in a vessel with flexible walls, as for example a partly blown-up balloon, then on heating the gas the molecules will travel faster and exert a greater force on the walls of the container. The gas will thus expand. If the walls are not flexible the gas obviously cannot expand, but the pressure it exerts on the walls will increase.

Applications of expansion

There are many applications of the expansion of substances when heated, as for example, thermometers, bi-metallic strips, and the fastening of iron tyres to railway wheels. The use of the thermometer has been discussed in the experiments and will not be considered further.

The use of bi-metallic strips depends upon the fact that two identical pieces of different metals will not expand the same amount, when heated through the same temperature range. Suppose then that these two metals, for example copper and iron, are fastened together. Since they are fastened together they must both be heated through the same temperature range but the copper will expand more than the iron. This is illustrated in the diagram below.

This bending of the two pieces of metal is used in thermostats when the temperature is used to control the electrical current flowing through a circuit. Suppose that at 20°C the two metals are the same length, i.e. the bi-metallic strip will be straight.

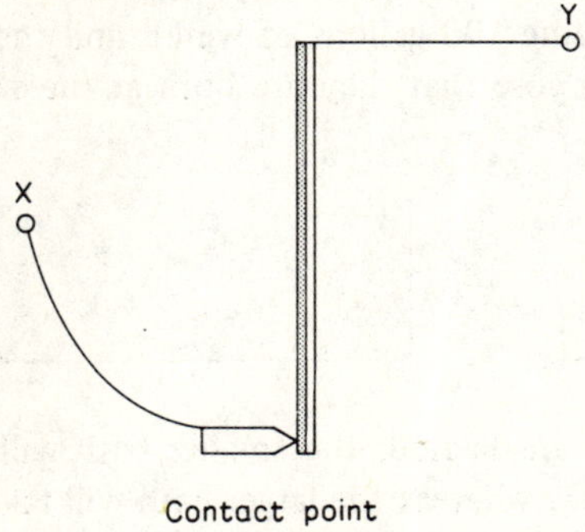

The electric current flows in at X, through the contact point, through the bi-metallic strip and out at Y, i.e. there is a complete circuit. If the temperature rises to 21°C, the bi-metallic strip will bend, the contact point will no longer touch the strip, the circuit will no longer be complete and thus no current will flow.

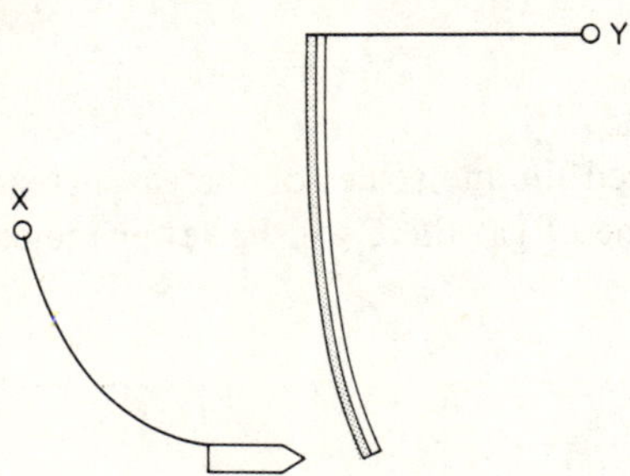

The current will not flow until the temperature of the bi-metallic strip has fallen to 20°C when the strips will again become straight and touch the contact point.

Iron tyres can be fitted to railway wheels, as many years ago they were to horse-drawn carts, by heating the tyre and causing it to expand, then placing it over the wheel and allowing it to cool and thus contract onto the wheel. In this way a very tight fitting metal tyre is produced.

Difference between Heat and Temperature

There is a great deal of confusion on the subject of the difference between heat and temperature. Temperature is merely a number saying how hot a body is. It does not give an indication of the amount of heat contained by the body. This may be seen clearly by considering two baths, one holding 100 gallons of water, and the other holding 1 gallon of water. Suppose that they are both at the same temperature, i.e. 20°C.

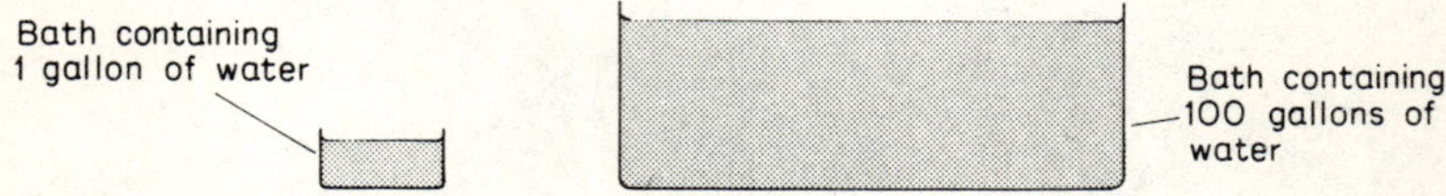

If the two baths are heated, the smaller bath will quickly attain a temperature of 25°C., whereas the larger bath will take much longer to attain this same temperature. Ultimately they will both be at 25°C, i.e. the same temperature, but the larger bath will contain more heat energy. Thus the temperature does not tell us how much heat the baths contain, but merely indicates that they are both as hot as one another, i.e. temperature is the hotness of the body.

Transfer of Heat

When a gas is heated the molecules of the gas increase in speed, and so in any given volume of gas there will be fewer molecules after heating than before.

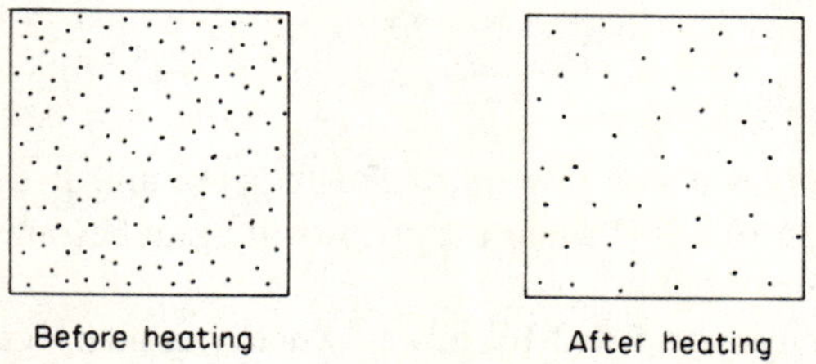

This has the effect of making that particular volume of the gas *lighter* after heating than before, i.e. the density of the gas has decreased.

Cold air is thus denser — heavier for the same volume — than hot air, and so will sink through the warm air, effectively pushing warm air upwards. Warm air thus rises above the cold air and the movement of this air produces the convection currents. The heat is consequently carried along by the body itself moving.

This procedure is almost the same for liquids except that the speed of the convection currents is lower than in gases.

Unlike a liquid or a gas the molecules of a solid cannot move freely through the solid, but are restricted to vibrating about the fixed lattice points. This means that convection currents can only occur in liquids and gases and not in solids.

When a solid is heated the molecules vibrate faster. In vibrating faster they will cause their neighbours to vibrate faster and so there is a general speeding up of vibrations of the molecules through the solid. In this way the heat energy is passed through the solid. The process is called *conduction,* and is defined as the passage of heat through a substance without the substance moving.

When the molecules of the substances are close together conduction will take place quite readily since adjacent molecules will readily interact with one another. Inevitably, as the molecules become further and further apart, the interactions between adjacent molecules becomes less and less, and so less and less heat is conducted through the substance. Since the molecules of a liquid are further apart than a solid and those of a gas further apart still, it follows that gases are the poorest conductors of all, followed by liquids, with solids — in particular metals — being the best of all.

The third method of heat transfer is by radiation. The main source of heat transferred to the earth is from the sun. This heat has to pass through a vacuum and therefore cannot be transferred by conduction or convection. This process of heat transfer is in fact by radiation. When transferred in this manner, the heat travels at the speed of light and can pass through a vacuum and transparent objects. Radiant heat is in fact similar to light and like light can be reflected by shiny surfaces.

The Vacuum Flask

Vacuum flasks are designed to keep substances either hot or cold. The main objective is to prevent heat entering or leaving.

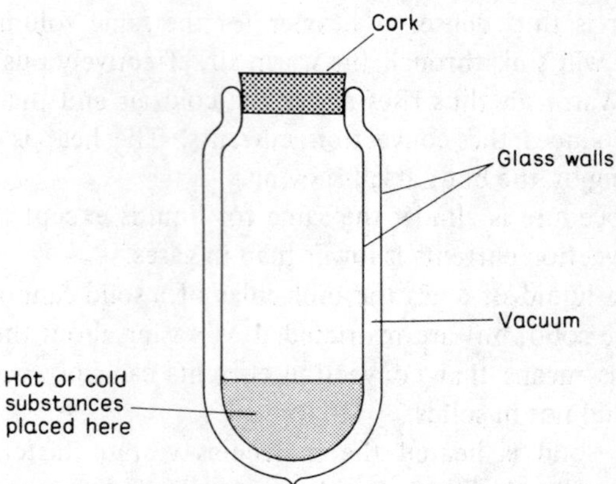

A vacuum flask is a double-walled glass container, the inside of both walls being silvered as in the diagram, with the space between them a vacuum.

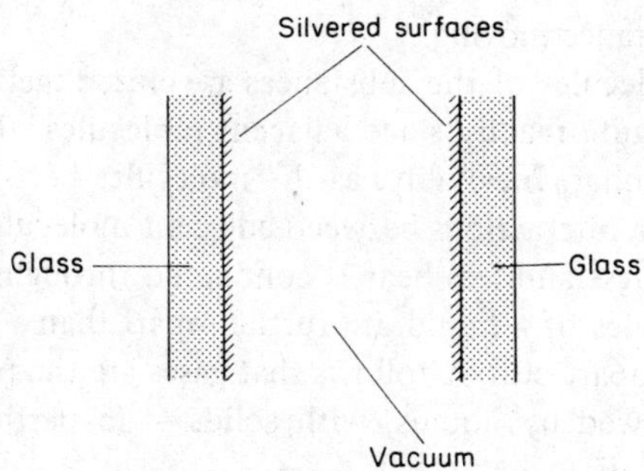

Heat transfer from the inside to the outside or *vice versa* is prevented in the following manner.

Convection

No convection occurs in the vacuum, thus heat can only be lost by convection through the glass walls or cork stopper. Since convection does not occur in solids no heat is lost by convection.

Conduction

Heat can enter or leave the vacuum flask by conduction through the glass and the cork but not through the vacuum. Since glass and cork are both poor conductors very little heat is gained or lost by conduction.

Radiation

Radiant heat can pass through the vacuum but the vast majority of it is reflected by the silvered glass surfaces.

Heat obviously does enter and leave the vacuum flask, and so the design is such as to make these losses very small. This in fact is what is observed in practice.

Chapter Three
Light

Section One

Experiment 28
The production of coloured light

Objectives
a To establish why certain surfaces appear coloured.
b To split white light into its basic colours.
c To produce a rainbow.

Apparatus per group
Red, blue, and green sources of light; red, blue and green surfaces: glass prism, or small mirror and flat dish; clear, plain, cylindrical glass beaker.

Procedure
i This first experiment is an attempt to establish why certain surfaces appear coloured and is probably best achieved by the teacher carrying out a series of small demonstration experiments, each experiment being followed by a class discussion.

Ideally the experiments should be conducted in a fully darkened room since white light from outside will complicate the results. If, however, the room can only be partially blacked out, then this will have to do.

ii The first series of experiments are to look at white objects under red light, blue light and then green light — use either flash lights with coloured filters in them, or coloured mains bulbs. The results obtained should be that in each case the surface no longer appears white but takes on the colour of the light shining upon it. This means that the white surface must be reflecting back all the light that falls upon it.

iii The second series of experiments are to look at a red surface under red light, blue light and then green light. You should find that when red light is shone onto the red surface, the surface appears red, but for the blue and green light it appears black. Now black is essentially the absence of light and hence it would appear that the red surface has absorbed the blue and green light, i.e. it absorbs all colours except red which it reflects.

iv These series of experiments should be repeated with green and blue surfaces viewed under the three different lights.

v In general we should be able to conclude that a surface appears coloured because it absorbs all other colours and reflects the particular colour that we see. For a surface to appear red, it means that red light from that surface must have passed into the viewer's eye and therefore cannot have been absorbed by the surface.

vi In a sense, this first series of experiments have inferred that white light, i.e. light from the sun, is made up of different colours, and this can be shown quite easily by splitting the white light up into its basic colours. This experiment can be performed by the children, but the teacher may find it easier to demonstrate it first.

vii There are two possible ways of breaking white light into its basic colours — these basic colours are called the *spectrum*. First of all by shining white light through a glass prism, or alternatively by shining light through a water prism.

viii If a glass prism is available a narrow beam of sunlight should be obtained by placing a card with a hole in it in front of the sunlight, which needs to be very strong. If the day is not very sunny, a substitute could be a fairly intense electric light. Let the light from the hole pass through the prism and observe the resulting spectrum on the walls or ceiling, as in the diagram.

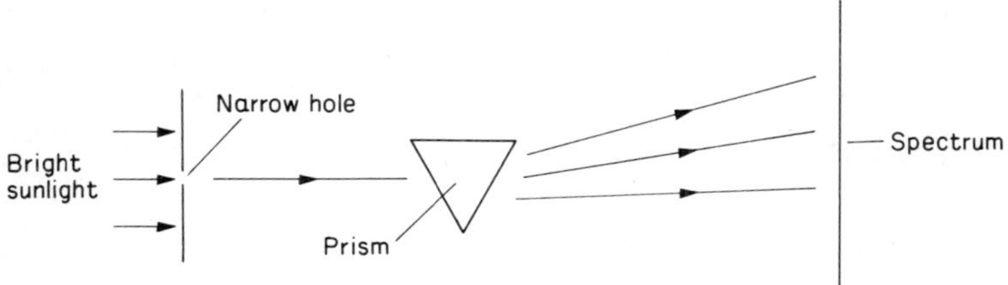

ix A second method, is to make a water prism. This can be done by putting about an inch of water in a dish or tray and leaning a small mirror against an inside edge. Arrange the dish so that the sunlight falls on the surface of the water. It will then pass through the water, be reflected by the mirror and pass out through the water.

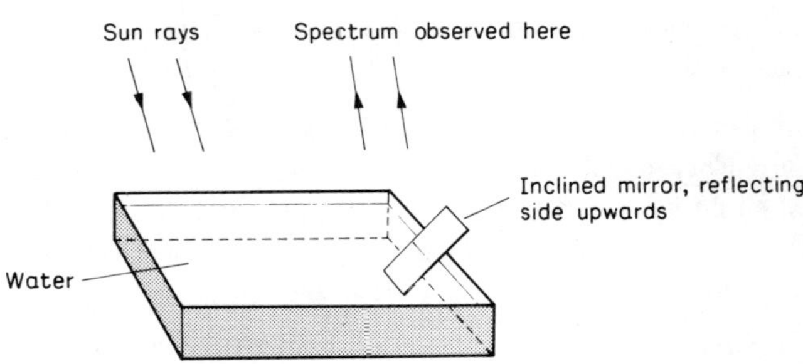

The spectrum can be observed on the ceiling or on a piece of white paper.

The colours of the spectrum are Red, Orange, Yellow, Green, Blue, Indigo, Violet.

x A rainbow is formed when sunlight passes through raindrops. Each raindrop acts as a prism splitting the sunlight into the colours of the spectrum.

There are two possible ways of producing a rainbow; the first method is to stand a clear, plain, cylindrical glass beaker full of water in bright sunlight about 1 metre above the floor. Make sure that the bottom of the beaker is not completely covered i.e. it needs to be stood on a surface as indicated in the diagram.

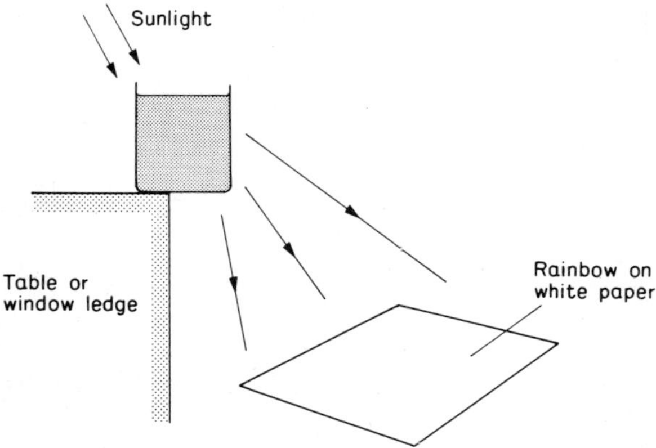

A second method of producing a rainbow is to go outside on a bright sunny day and stand with your back to the sun, facing a reasonably dark background. Spray water from a fine nozzle hose towards the dark background and you should observe a rainbow.

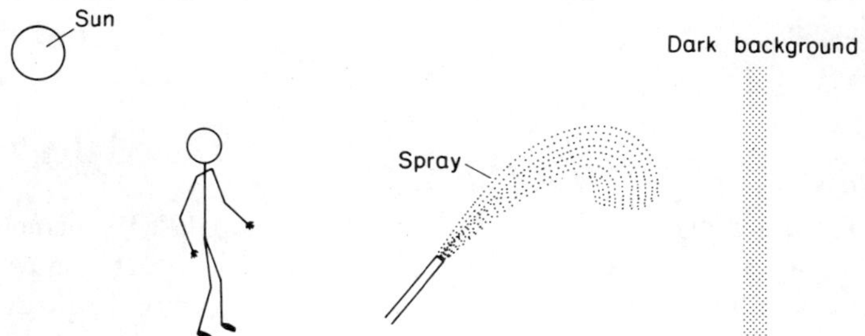

Notes

1 In the first series of experiments you may find, for example, that a red surface under blue or green light does not appear completely black. This is because the surface will not have been a pure red but will have contained traces of other colours.

2 A glass prism is quite expensive but is not really necessary since the water prism is an acceptable substitute.

3 The reason why a rainbow is actually part of a circle is difficult to explain and explanations should be avoided.

Experiment 29
To mix coloured lights

Objectives

a To show that mixing the seven colours of the spectrum produces white light.

b To show that it is possible to produce white light from *three* of these colours, called the primary colours.

c To produce three secondary colours from the primary colours.

d To produce white light from a primary and secondary colour.

Apparatus per group

Selection of discs, paints corresponding to the seven spectrum colours, or seven flash-lamps and seven filters, one for each colour of the spectrum.

Procedure

i   There are two main methods of mixing coloured lights, namely by mixing light on a screen which has been produced from flash lamps with coloured filters placed in front of them (see apparatus construction, page 128), or by spinning discs or tops which have been coloured with the appropriate colours (see apparatus construction, page 128). In the descriptions that follow, the spinning discs method will be considered.

ii The class should make a series of discs to spin and should have a supply of paints corresponding to the spectrum colours.

iii A disc should be divided into 14 equal segments and the segments should be coloured as in the diagram.

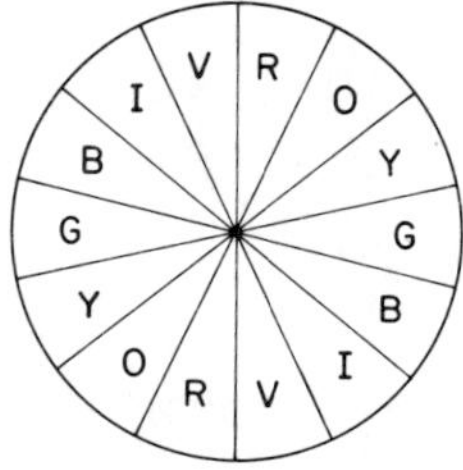

R — Red
O — Orange
Y — Yellow
G — Green
B — Blue
I — Indigo
V — Violet

The disc should then be spun as fast as possible and the resulting light seen from the disc should appear white.

iv Another disc should now be divided into six equal segments and using the three colours RED, GREEN and BLUE, the segments should be coloured as indicated.

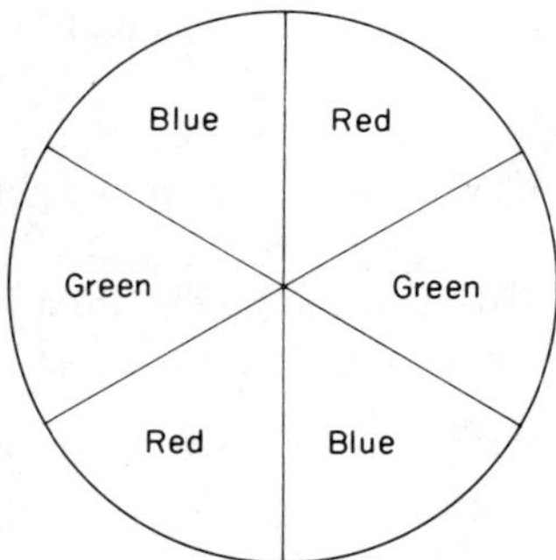

On spinning the disc the light appearing should be white, showing that in fact white light can be produced from these three colours. These colours are called the *primary light colours.*

v Another disc should now be divided into four equal segments and using two primary colours, e.g. red and blue, the segments should be coloured as indicated.

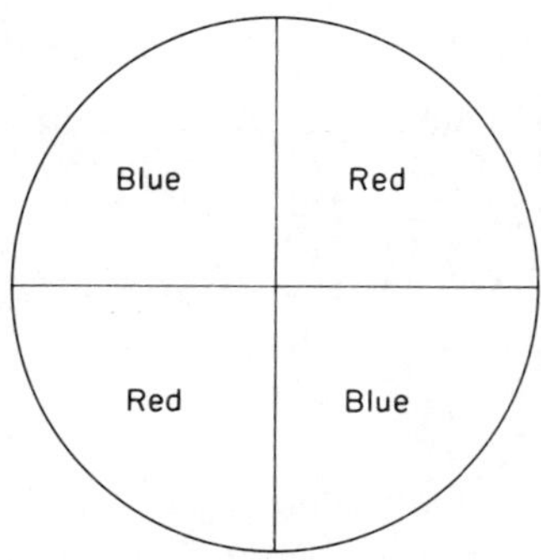

On spinning the disc the light appearing should be viewed and should appear to be purple — the correct name for it being *magenta.* Therefore: Red + Blue = Magenta.

vi This procedure should be repeated for red and green, and green and blue. The results obtained should be:
Red + Green = Yellow
Blue + Green = Peacock Blue

vii The three colours, magenta, yellow and peacock blue are called *secondary light colours.*

Since yellow is made from red and green light, and white light is made from red, green and blue light, then white light should be produced by yellow and blue light.

This can be illustrated by dividing a disc into six equal segments and using yellow and blue, colour the segments as indicated in the diagram.

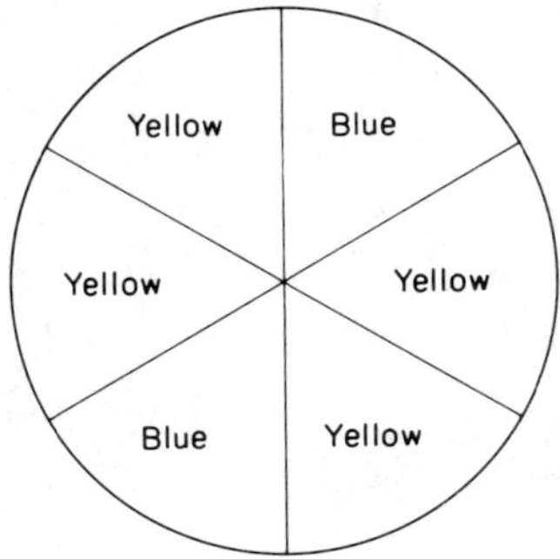

On spinning the disc you should produce white light.

viii   This procedure should be repeated for:

A Magenta and green.

B Peacock blue and red.

This shows that a secondary and the appropriate primary colour when added will produce white light.

Notes

1 The reason for describing spinning discs is that the method lends itself to class experiments rather better than using flash-lamps.

2 We are mixing coloured light and not coloured paints. In art the three primary colours are different and hence yellow paint is not produced by mixing red and green paint. This point should be emphasised if raised by the children.

3 When mixing a secondary and primary colour it is necessary to have twice as much of the secondary colour as of the primary colour (for example, four segments to two segments in the last diagram). In total this in fact gives us two segments of each of the three primary colours.

Experiment 30
Light travels in straight lines

Objectives
a To show that light travels in straight lines.
b To make a pin-hole camera.

Apparatus per group
Four mounted cards, length of cotton, or thin string, cardboard box approximately 30 x 15 x 10 cm., tissue or grease-proof paper.

Procedure
i The class should be issued with the mounted cards (see apparatus construction, page 129). They should then be asked to position the four cards as indicated in the diagram so that they can view an object through all four holes.

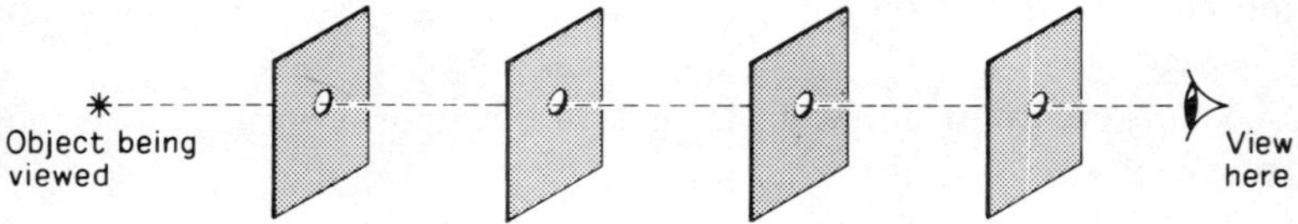

They should then be asked to pass a piece of cotton or thin string through all four holes without moving the cards, and to pull the cotton taut. They should observe that all four holes lie on a straight line.

The class should then be asked to move one of the cards slightly and see if they can still see the object.

In this way they should establish that the object can only be viewed when all four holes lie on a straight line. This means that the light from the object must be passing through all of these holes into the eye, i.e. light travels in straight lines.

ii An application of the fact that light travels in straight lines is the pin-hole camera and the following is a description of how one may be made.

Take an empty cardboard box, approximately 30 x 15 x 10 cm., for example a shoe box, and cut out a large square at one end. Cover the space with tissue or greaseproof paper as indicated in the diagram.

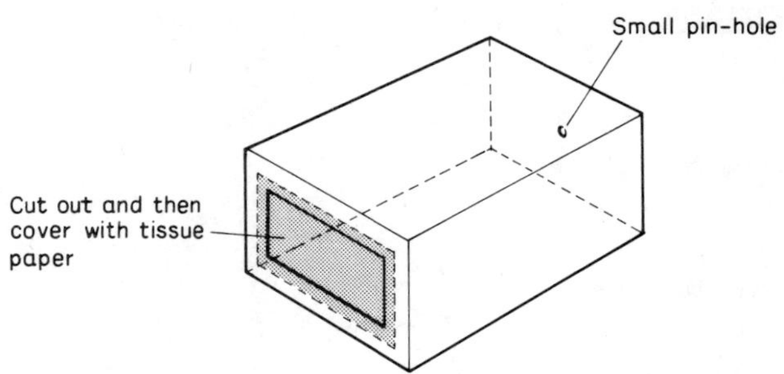

In the opposite end of the box make a small pin-hole and securely fasten the top or any loose sides — ideally the box should be light-proof other than through the small pin-hole and the tissue-papered end.

Hold the box with the pin-hole towards the window and cover your head and the tissue-papered end with a dark cloth. You should then see a picture of the window on the tissue paper. Do not hold your head too close to the tissue paper — about 1 metre away is the optimum distance.

You should notice that the image of the window is upside down and reduced in size.

The children should be encouraged to experiment by making more than one hole or by enlarging the original pin-hole.

A rather more refined pin-hole camera can be made from two boxes, so that one box slides inside the other. Make a hole in one of the boxes and have the tissue paper screen in the other, as in the diagram.

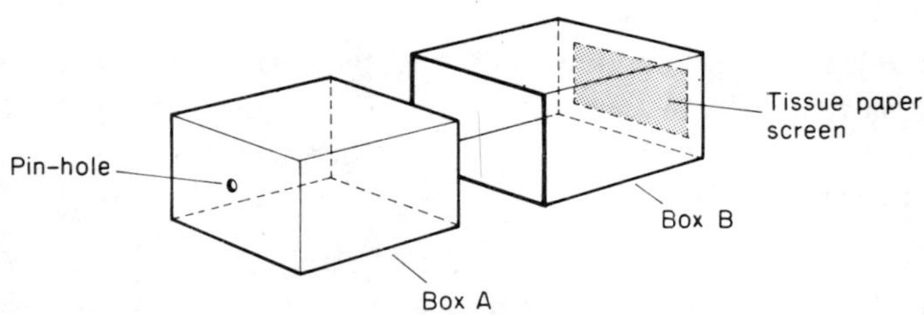

Box A just slides into Box B so that the pin hole and the tissue-paper screen are on opposite sides.

The size of the image can now be changed by sliding one box further into or out of the other.

Notes
1 A pin hole camera would really have a photographic plate instead of the tissue-paper screen.
2 In the refined model of the camera make sure that no light enters the box where the two boxes slide into each other.

Experiment 31
Reflection by a plane mirror

Objectives
a To show that a plane mirror reflects rays of light.
b To make a model periscope.
c To consider lateral inversion.
d To consider some applications of reflection.

Apparatus per group
Two plane mirrors, comb, flash-lamp, plasticine, long ruler or thin piece of wood, sheet of clean glass, jar of water, solid screen.

Procedure
i Issue each group with a flash light, a small plane mirror, a sheet of white paper and a comb.

Rays of light can be produced by standing the comb vertically on the white paper and shining the light at it, as in the diagram.

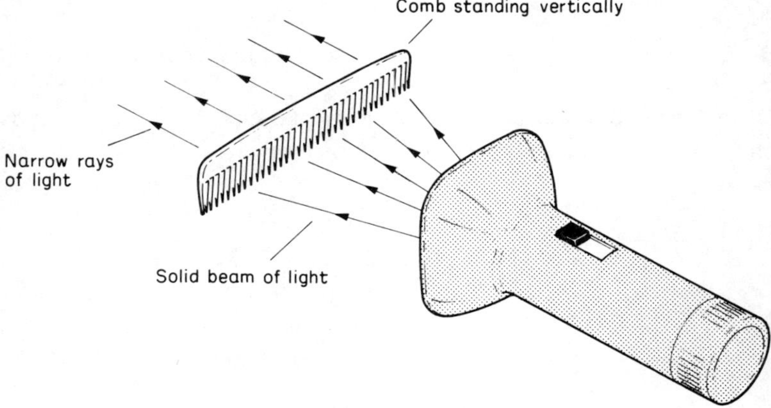

The children should then allow these thin rays of light to fall onto a plane mirror, where they will see them reflected. The diagram that follows is a plan view of the reflection.

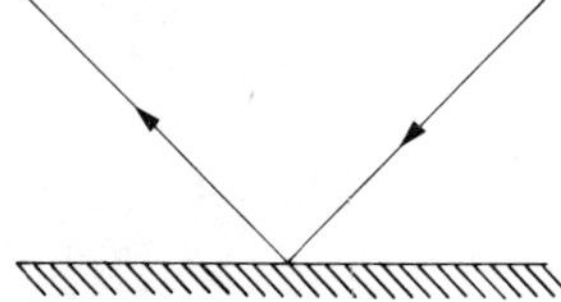

The diagram above is that used to represent a mirror. The reflection takes place from the silvered surface of the mirror, i.e. the rear surface of the glass, not from the front glass surface.

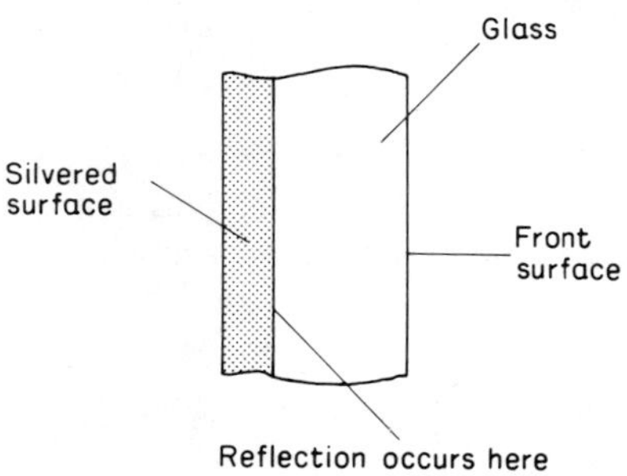

In the scientific diagram of a mirror the glass is never included, it is just the silvered surface that is drawn in.

ii Since light can be reflected it is possible to use mirrors to see around corners.

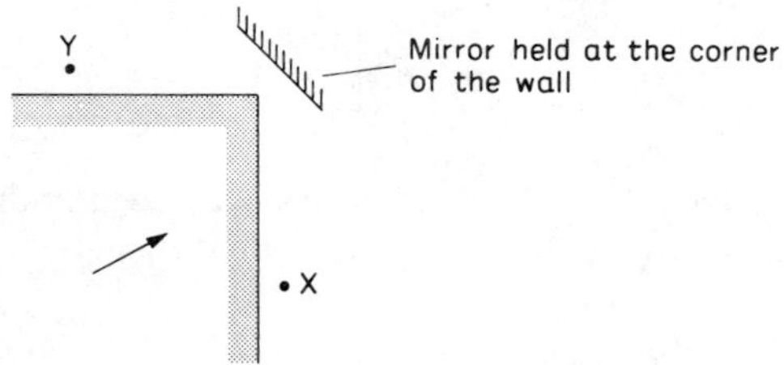

Let the child stand at X so that he cannot see Y around the corner. Ask him to hold the mirror so that he can see Y without moving. The position in which he will have to place the mirror is indicated in the diagram.

This procedure can now be modified to construct a model periscope.

Mount two mirrors in a ruler or long piece of wood using plasticine, as indicated in the diagram.

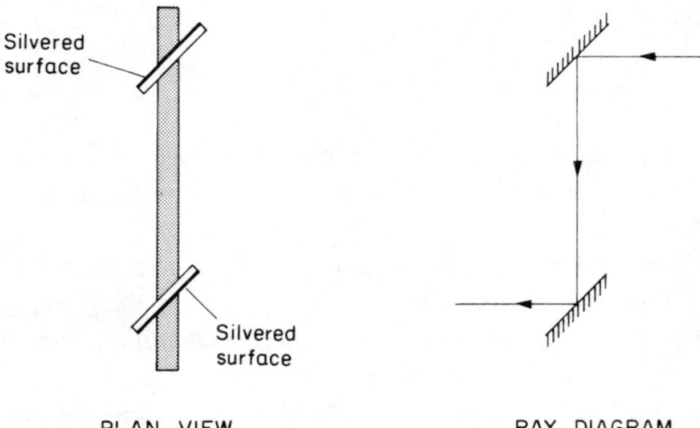

The first diagram is a plan view, whereas the second diagram is a scientific diagram showing what would happen to a single ray of light.

Using this arrangement it is possible for a child to sit on the floor and see something placed on a table — his head being below the level of the top of the table.

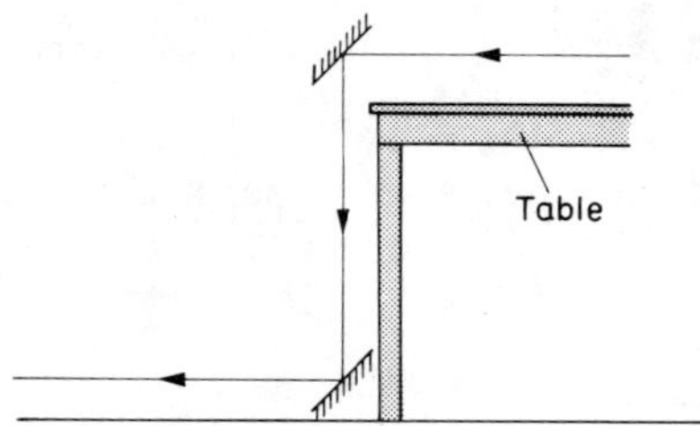

Alternatively the two mirrors can be arranged slightly differently and so enable someone to see things high up behind them.

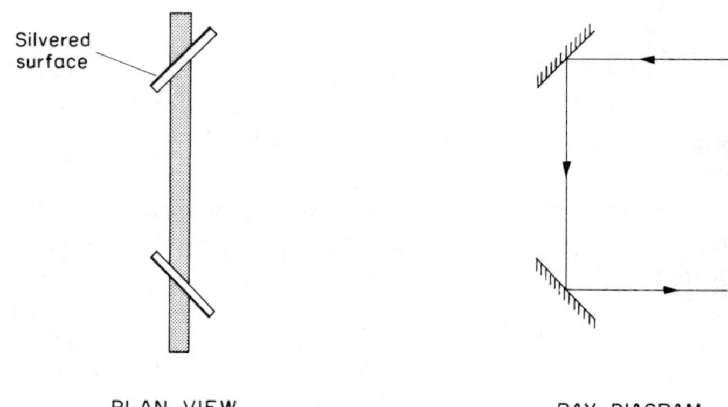

iii Ask a child to face a mirror and hold his *right* hand forward as though to shake hands with his image in the mirror. Ask him to say what he sees.

It appears that the image is offering his *left* hand. In fact investigation will show that the image is completely reversed, but it is not upside down. This type of inversion is called *lateral inversion*.

This can be very useful when trying to read writing on blotting paper. The children should use a piece of blotting paper on wet writing and then read the message on the blotting paper in the mirror.

Very often on the front of an ambulance, the word is laterally inverted so that a motorist looking in his driving mirror will read it the correct way round. The class should be asked to write any word laterally inverted and then check, by reading it in the mirror, that they were correct.

iv Two interesting applications of reflection are copying by reflection and Pepper's Ghost.

To copy by reflection set a piece of clean glass vertically, with the drawing to be copied on one side and a piece of white paper on the other. Look through the glass at the white paper and draw around the reflection of the drawing.

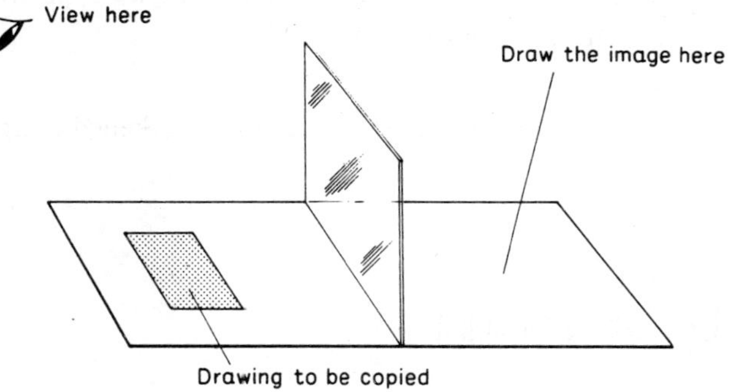

Pepper's Ghost is an old music-hall trick in which the audience apparently saw on the stage a ghostly figure through which people could walk. In fact what they saw was the image of a person standing in the wings of the theatre.

This can be illustrated by the following demonstration.

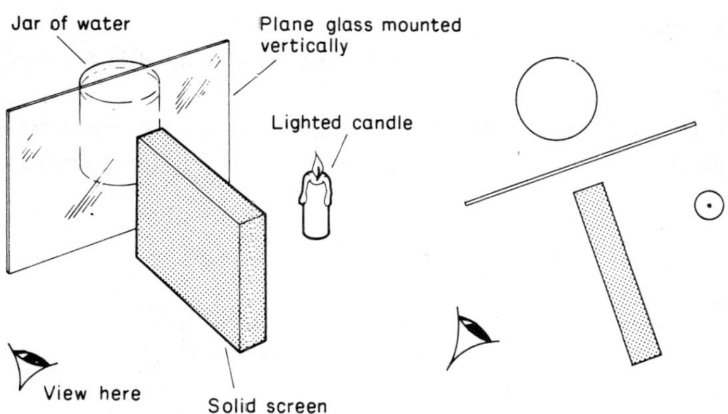

A plane sheet of glass is mounted vertically and a jar of water placed on the other side of the glass from the observer. A lighted candle is positioned as indicated with a solid screen placed between the candle and the observer. This means that the observer is unable to see the candle directly, but on looking through the glass he should see the image of it, and if everything is correctly positioned this image should be in the middle of the jar of water. Thus we apparently see a candle burning in a jar of water.

Notes
1   This last demonstration is most effective in a darkened room.

Experiment 32
The bending of light

Objectives
a   To show that light rays when passing from water into air or *vice-versa* are bent.
b   To illustrate the working of a lens.

Apparatus per group
Flash-lamp, comb, flat-sided bottle, jam jar, magnifying glass.

Procedure
i   The first experiment is probably best demonstrated in front of the class by one of the children.

Place a coin in the middle of an empty sink and ask one of the class to stand so that the coin is just visible, as indicated in the diagram.

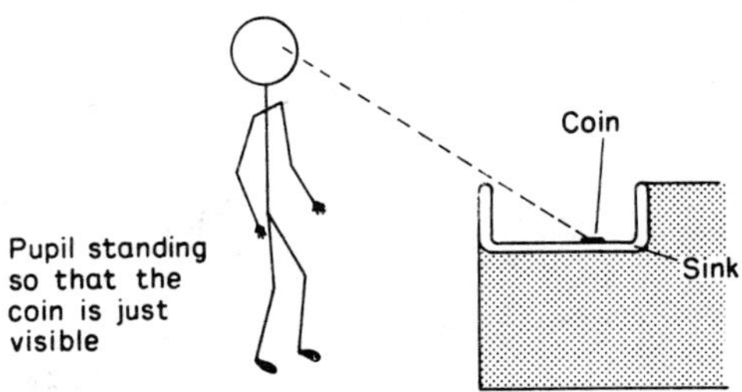

The pupil should then be asked to move one step backwards so that the coin is no longer visible, and the class should be asked how the coin could be made visible without moving it or the pupil. One method could be using a mirror, i.e. bending the rays of lights coming from the coin and reflecting them into the pupil's eyes.

Another method is gradually to fill the sink with water, making sure that the position of the coin is not disturbed.

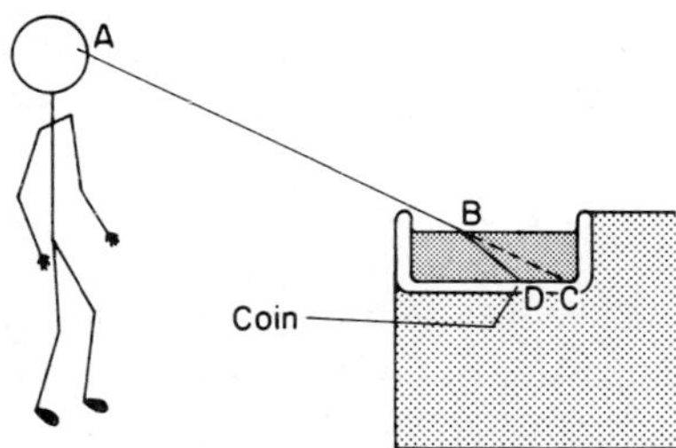

Without the water, in the sink, anything to the left of the point C is not visible, i.e. ABC is a straight line, and light travels in straight lines. When the water is added, the coin becomes visible, and thus the rays of light from the coin to the pupil's eyes must have followed the path DBA. Since the light travels in straight lines in the air and the water, it follows that its path must have been bent on leaving the water and passing into the air.

ii This bending can be viewed directly by performing the following experiment.

Issue each group with a flash-light, a comb and a flat-sided bottle.

The flash-light should be shone through the comb to produce a selection of narrow rays of light. For this experiment it may be advantageous to use pieces of card attached to the comb to reduce the number of rays to one or two. The flat-sided bottle should be filled with water and placed on its largest surface on a piece of white paper. The rays of light from the comb should then be allowed to pass through the bottle of water making sure that initially they enter the water at an angle, as indicated in the diagram.

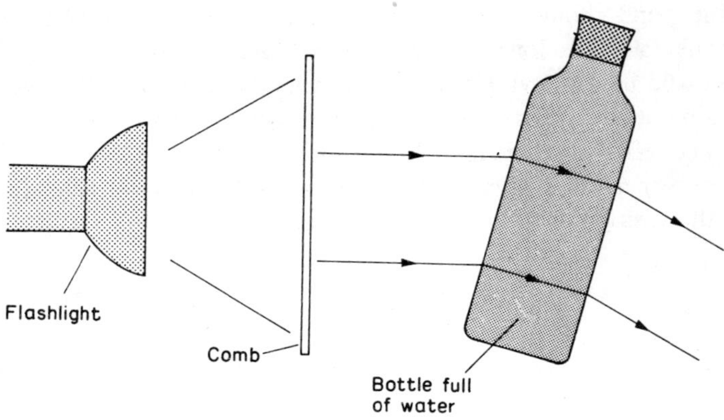

The rays of light will be seen to be bent on passing into the water and then bent again when passing out of the water. It should be noticed that the bending only occurs when the light passes from one substance to another, not while actually passing through the substance.

iii This phenomena of bending can be further illustrated if the class look at a straight stick or pencil standing at an angle in a beaker of water. They should set this up and describe what they see.

iv The action of a lens can easily be seen by using a 'water lens'. Replace the bottle full of water which was used in the earlier experiment, with a jam-jar full of water, and shine the light through it as indicated in the diagram.

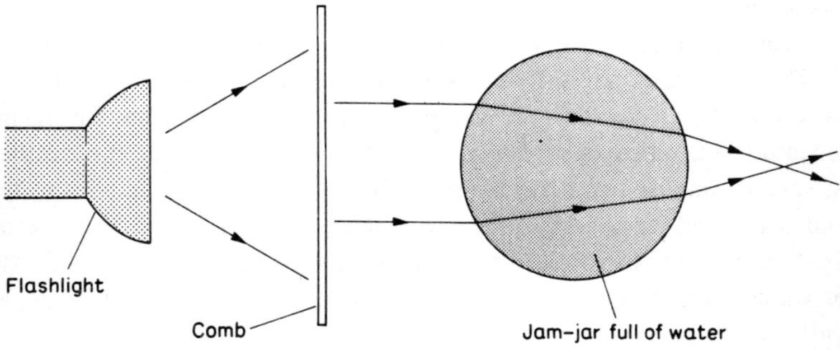

The rays of light bend and come together to a point. This lens has *converged* the light and is said to be a converging lens or a *convex lens*. This type of lens is always fatter in the centre than at the edges and is always drawn thus:

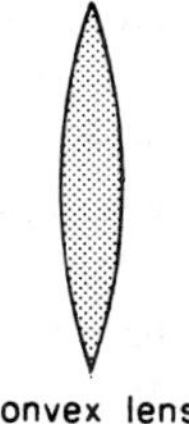

Convex lens

Some lenses, however, will diverge the light, i.e. spread it out, and these are called *concave lenses*. This type of lens is always thinner in the centre than at the edges and is always drawn thus.

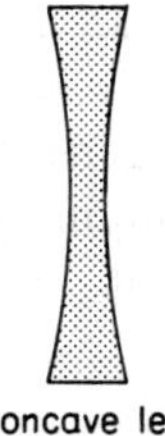

Concave lens

v Some of the properties of convex lenses can be investigated using magnifying glasses, since these are always convex.

Using the magnifying glass, the class should try and form an image of the window on a sheet of white paper. They should stand about 10 metres away from the window and hold the paper reasonably near to the lens as in the diagram.

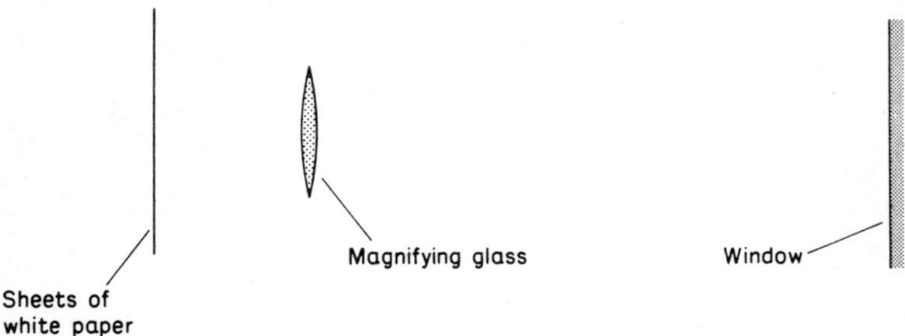

The class should then be asked to describe the image — which will be reduced in size and upside down — and measure the distance between the lens and the sheet of white paper.

If it is a bright sunny day, the class should focus the sun's rays onto a sheet of paper and make it start to burn. They should be asked to measure the distance between the lens and the paper, and should find that it is the same as in the previous experiment, showing that the sharp spot produced on the paper is the image of the sun.

The class should be told to stand about 10 metres away from the window, and hold the magnifying glass about 1 metre in front of them so that they are looking through it at the window. They should be asked to describe what they see — the image of the window should be smaller than the window and upside down.

They should now walk slowly towards the window looking through the magnifying glass all the time, and continually describing what they see. The image should be getting larger but still upside down.

Eventually they should reach a point when the image goes blurred and then suddenly it will go the correct way up and become magnified. This shows that to act as a magnifying glass, the lens must be held quite close to the object being magnified.

Notes

1 When shining rays of light through the flat-sided bottle and the jam-jar it may be helpful to add a drop or two of milk to the water to make the rays of light more visible.

2 When using the magnifying glass, do not allow the children to look at the sun or a bright light through it, since if they focus the rays sharply onto the retina of their eyes it could cause damage.

Section Two

Apparatus required and suppliers

The items listed below are sufficient to make one complete set of apparatus for the experiments indicated.

Experiments 28, 29, 30, 31 and 32

Sources of coloured light (see apparatus construction, page 128).
Red, blue and green surfaces, e.g. painted sheets of paper.
Small glass prism.
Two small flat mirrors — approximately 10 x 10 cm.
A flat dish — approximately 30 x 30 x 5 cm.
A clear, plain, cylindrical glass beaker, approximately 10 cm. in diameter and 15 cm. high.
Paints corresponding to the seven pure colours of the spectrum.
Selection of discs, (see apparatus construction, page 128).
Four mounted cards (see apparatus construction, page 129).
Cardboard box, approximately 30 x 15 x 10 cm.
Tissue or grease-proof paper.
Flash-light.
Comb.
Plasticine.
Long ruler.
Sheet of clean glass, approximately 50 x 50 cm.
Magnifying glass.
A flat-sided medicine bottle.

All of these items can be obtained easily from a chemist or hard-ware shop with the exception of the small glass prism which will probably have to be purchased from a commercial suppliers such as either of the following:

Philip Harris, Frederick Street, Birmingham B1 3DJ

Griffin & George Ltd., Ealing Road, Alperton, Wembley, Middlesex HA0 1HJ

Section Three

Apparatus construction

Sources of coloured light

There are two main methods of producing coloured light. The first is to use a mains electric light bulb that has been painted — these are sold commercially; the second is to replace the glass in the front of a flash-light by a piece of coloured plastic or glass. If in the later experiments you are going to mix light using the spinning discs you would be well advised to use the coloured mains lights to produce the coloured light. In this case choose a low wattage bulb, of the order of 60 watts. If the wattage is too high the coloured light will contain too much white light and so obscure the results of the experiment.

Spinning Discs (for Experiment 29).

Cut out circles of thin cardboard, approximately 15 cm. in diameter. Divide the disc into the appropriate number of segments, colouring each segment as described in the experiment. Make a hole in the centre of the disc and push through a sharp pencil about 10 cm. long, as in the diagram. The disc can then be spun on the pencil point.

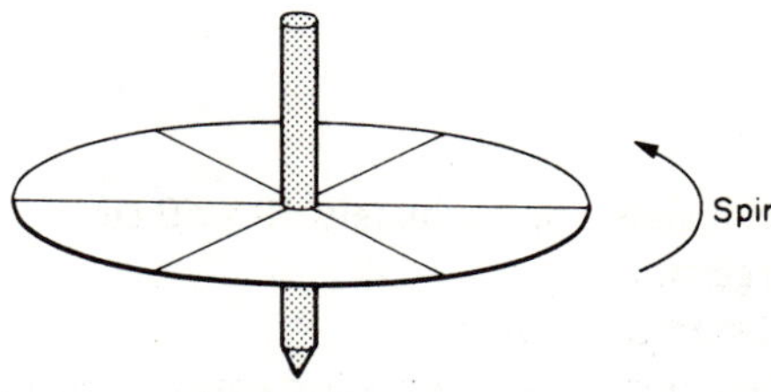

Mounted Cards (for Experiment 30).

Cut four pieces of stiff cardboard approximately 15 cm. square and fasten them, with drawing pins, to small wooden blocks so that they will stand upright. Very carefully make a small hole through each card at exactly the same height above the surface they are standing upon. In this way the four holes can be set in a straight line to allow an object to be viewed through them.

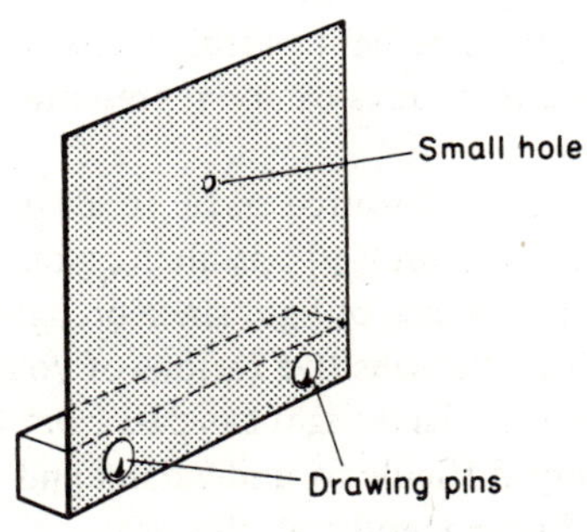

Section Four

Additional background information

Sources of Light

Any hot body, for example the sun, a candle or an electric light bulb, is a source of light. The light passing from these hot bodies is said to travel in rays, and these rays can be represented on diagrams by straight lines.

Rays of light from a source travel in all directions, and only those which enter your eyes are responsible for your seeing the object. Darkness is simply the absence of light since no light rays have entered your eyes and stimulated the sensitive regions of your retina.

The reasons why and how light can pass from the sun to earth through a vacuum are very difficult to understand and cannot be described in a simple theory. Consequently at this stage it is better to concentrate upon the properties exhibited by light and light rays rather than worrying about what light actually is.

Coloured Bodies

Objects such as a book or a house do not produce light and yet can be seen. This can be illustrated by taking a book into a darkened room. While the light is on, the book can be seen, but as soon as the light is switched off the book cannot be seen, thus showing that the book does not produce light.

The reason that the book can be seen is that light rays from a source of light, in this case the electric light bulb, fall onto the book and are reflected from the book in all directions. Some of these reflected rays will pass into the observer's eyes and so the image of the book is formed.

If an object appears red it is because that object has absorbed all the other colours of the spectrum and reflected the red light. The different shades of red are because the object may not have absorbed quite all, for example, of the blue light, and the reflected red light will contain a little blue light as well. A black object has absorbed all light and reflected none, whereas a white object has absorbed no light and reflected it all.

Mixing of Coloured Light

White light is composed of the seven basic colours, red, orange, yellow, green, blue, indigo and violet. It can, however, be made up from the three primary colours, red, blue and green. The mixing of these three primary colours can be represented quite easily on the following diagram.

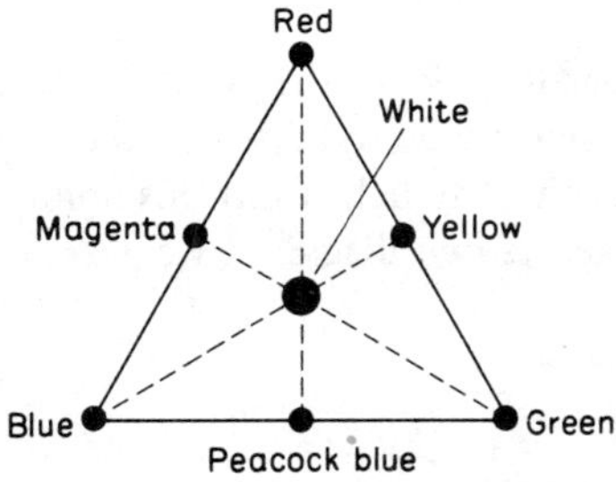

There is a difference between the three primary colours in light and those in art, and it must not be imagined that the results of mixing coloured paints will be the same as those of mixing coloured lights.

Light travels in straight lines

The fact that rays of light can be considered to be straight lines, i.e. light travels in straight lines, is very important, since without this phenomena we would not be able to see objects at all clearly, we would not be able to judge distances with any degree of certainty, and such things as sharp shadows would not exist.

It is possible, using the fact that light travels in straight lines, to discuss and explain eclipses. The case being considered is when the moon passes between the sun and the earth.

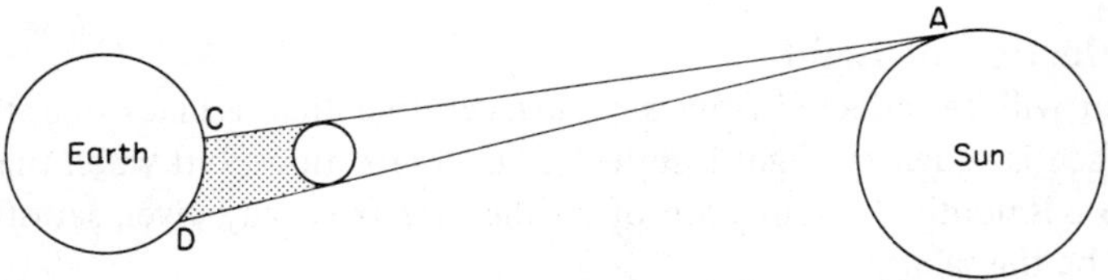

If we consider light from the edge of the sun, i.e. point A in the diagram above, falling onto the earth, we see that no light from this point falls on the region CD, but it will fall on all other regions.

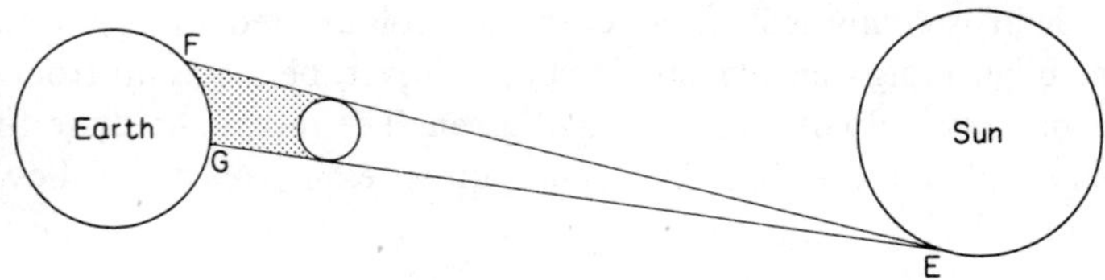

If we now consider light from the point E, i.e. the extreme point of the surface directly opposite to A, falling on the earth, we have a region FG which will not receive any light from this point.

If we now combine the two diagrams, we have:

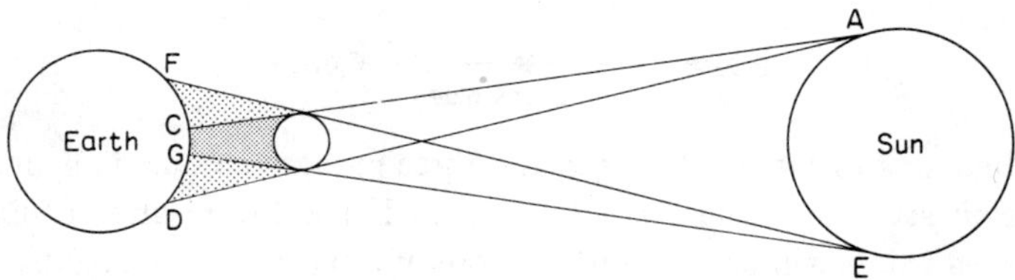

We thus have the region CG which will receive no light at all from the sun. This region is called the *umbra,* and is the region of total eclipse. The regions FC and GD will not receive the full amount of light from the sun but will receive some, and they are thus called the *penumbra.* These are the regions in which partial eclipses occur. The regions beyond F and D will receive sunlight from the whole of the sun and will not show any evidence at all of the eclipse.

Reflection of Light

Light will be reflected from any surface other than a black one. If the surface is rough the light is reflected in all directions, but when the surface is smooth the reflection of all the rays from any given set of rays will be the same.

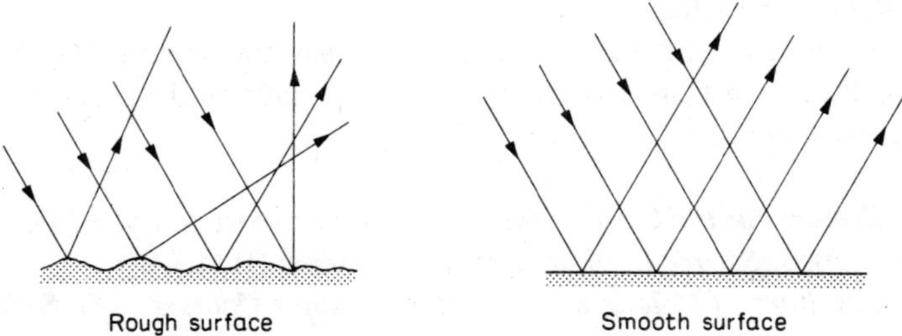

It is because of this regularity of reflection from smooth surfaces that images can be formed.

Bending of Light

When light passes from one medium into another, for example air into glass, or air into water, it is deviated or bent. This process is called *refraction,* and can be represented diagramatically thus:

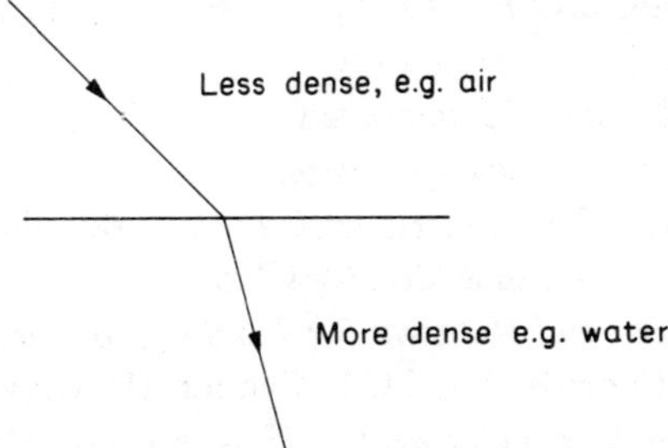

The bending or deviating only occurs at the surface of the two media.

Additional Reading

The following is a list of books the reader may find useful. It is not a comprehensive list and no attempt has been made to list them in any particular order.

1. *Science for the C.S.E. Year* – B.K. Strong (Arnold and Son).
2. *Unesco: Source Book for Science Teaching* (H.M.S.O.).
3. *Teaching of Colour in Elementary Science Courses* – G. Savage (John Murray).
4. *Introduction to Science* – K.G. Brocklehurst and F. Winterbottom (English Universities Press).
5. *Science Workbooks* – M.E.J. Shewell and R.W. Crossland (Heinemann Educational Books).
6. *An Approach to Primary Science* – S. Redman, A. Brereton and P. Boyers (Macmillan).
7. *Approaches to Science in the Primary School* – C. Lawrence (Ward Lock Educational).
8. *Magnets and Electricity* – G.E. Allen (Education Supply Association).
9. *The Book of Experiments* – L. de Vries (John Murray).
10. *Simple Science Experiments* – A. James (Schofield and Sims).
11. *Science Experiments with Inexpensive Equipment* – C.J. Lynde (Van Nostrand & Co., New York).
12. *Methods and Materials for Teaching General and Physical Science* – J.S. Richardson and G.P. Cahoon (McGraw-Hill).
13. *Simple Science Projects* – R.A. Swallow (Nelson).
14. *Everyday Science Topics* – T.A. Tweddle (Harrap).
15. *Elementary Physics* – W. Littler (Bell).
16. *The Tricks of Light and Colour* – H. McKay (Oxford University Press).
17. *Discovering Science Series* – D.H. Barratt (Arnold).
18. *Working With Series* – E.A. Catherall and P.N. Holt (Bailey Bros. and Swinfen).
19. *Junior Science Topics* – W.G. Western (Oxford University Press).
20. Reference book: *Science for Primary Schools, Part 2.* (John Murray, for the Association for Science Education). List of Books – Association for Science Education.